MW01702625

STECK-VAUGHN FUNDAMENTAL SKILLS FOR READING: READING COMPREHENSION AND REFERENCE SKILLS BEGINNING

Photo Credits

3 *digital clock* ©Houghton Mifflin Harcourt; 3 *clock* ©Hal_P/Shutterstock; 4 *clock* ©Hal_P/Shutterstock; 5 *clock* ©Hal_P/Shutterstock; 7 *calendar* ©C Squared Studios/Getty Images; 9 *invitation* ©Getty Images; 12 *sock* ©Houghton Mifflin Harcourt; 16 *Hermann Park map* ©Houghton Mifflin Harcourt; 16 *US map* ©Houghton Mifflin Harcourt; 16 *battle map* ©Houghton Mifflin Harcourt; 17 *floor map* ©Houghton Mifflin Harcourt; 17 *Hawaiian map* ©Houghton Mifflin Harcourt; 18 *Venn diagram* ©Houghton Mifflin Harcourt; 19 *eye diagram* ©Houghton Mifflin Harcourt; 22 *calendar* ©Shutterstock; 22 *clock* ©Hal_P/Shutterstock; 24 *map* ©Houghton Mifflin Harcourt; 25 *bar graph* ©Houghton Mifflin Harcourt; 27 *Chihuahua* ©Houghton Mifflin Harcourt; 34 *definitions* ©Houghton Mifflin Harcourt; 35 *glossary* ©Houghton Mifflin Harcourt; 67 *butterfly* ©Scott Slattery/iStockPhoto.com; 69 *tractor* ©C. Borland/ Getty Images; 69 *farmer* ©Getty Images; 77 *deer crossing* ©Corbis; 77 *USB cable* ©Getty Images; 77 *escalator sign* ©Getty Images; 77 *no smoking* ©Getty Images; 77 *construction sign* ©Getty Images; 77 *railroad sign* ©Houghton Mifflin Harcourt; 78 *railroad sign* ©Houghton Mifflin Harcourt; 78 *deer crossing* ©Corbis; 78 *poison* ©Getty Images; 78 *escalator sign* ©Getty Images; 78 *no smoking* ©Getty Images; 78 *construction sign* ©Getty Images; 79 *food label* ©Ilene MacDonald/Alamy; 80 *hand washing* ©Shutterstock; 81 *fire instructions* ©Houghton Mifflin Harcourt; 83 *dental chair* ©Gergely Zsolnai/ Shutterstock; 86 *computer* ©Getty Images; 88 *nurse and patient* ©Corbis; 88 *nurse* ©michaeljung/Fotolia; 89 *milk* ©Houghton Mifflin Harcourt; 90 *cows* ©Wally Stemberger/Shutterstock; 91 *grazing cattle* ©Alamy; 94 *living room* ©Getty Images; 94 *garage apartment* ©Houghton Mifflin Harcourt; 94 *cat* ©Getty Images; 94 *happy face* ©Getty Images; 94 *house* ©Library of Congress Prints & Photographs Division; 94 *dog* ©Getty Images; 94 *no smoking* ©Getty Images; 94 *apartments* ©Jorge Salcedo/Shutterstock; 97 *plot diagram* ©Houghton Mifflin Harcourt; 99 *plot diagram* ©Houghton Mifflin Harcourt; 111 *plot diagram* ©Houghton Mifflin Harcourt

Printed in the U.S.A.

ISBN 978-0-544-79072-8

1 2 3 4 5 6 7 8 9 10 0918 24 23 22 21 20 19 18 17 16

4500577711 A B C D E F G

TABLE OF CONTENTS

LESSON 1: Money

People spend and earn **money** every day. They compare prices and shop for products and services.

Here are some symbols often used to show amounts of money.

Symbol	Meaning	Example of use
$	dollar	Movie tickets cost **$**10 for each adult.
¢	cent—one penny is one cent	When your grandfather was young, he could buy a movie ticket for 10**¢**.
.	decimal point—separates dollars and cents	I paid $4**.**50 for a bag of popcorn at the movie theater.

When you read money written in numbers and symbols, look for the decimal point (.). The numbers to the left of the decimal point are dollars. The numbers to the right of the decimal points are cents. Read the numbers to the left of the decimal place first. Then read the numbers to the right. The decimal point is read as *and*.

Examples:

Jane has $126.80. (Read: *Jane has one hundred twenty-six dollars and eighty cents.*)
She saved $450.82 this month. (Read: *She has saved four hundred fifty dollars and eighty-two cents this month.*)

Money amounts can be written in several ways. Look at the standard form and word form. You may see money amounts in either form when you read.

Standard Form	Word Form
$10.12	ten dollars and twelve cents
$4.50	four dollars and fifty cents
$109.18	one hundred nine dollars and eighteen cents
$23.79	twenty-three dollars and seventy-nine cents
$126.00 or $126	one hundred-twenty six dollars

Knowing how to read amounts of money in both standard form and word form is an important skill to have. When people write a check or money order, they use words, numbers, and symbols. You need to be able to read a check to make sure you are paid the right amount of money.

Example:

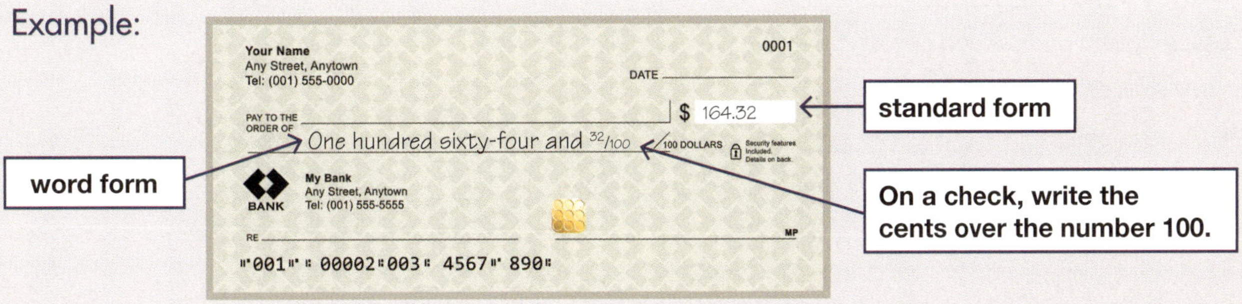

Dollar amounts with four or more numbers to the left of the decimal need commas. The comma separates every three numbers to the left of the decimal point.

Standard Form	Word Form
$5,900.00	five thousand nine hundred dollars
$12,050.00	twelve thousand fifty dollars
$440,000.00	four hundred forty thousand dollars

You may see very large dollar amounts written in different ways when you read.

Examples:

Last year, the company invested $1,200,000 in new building projects.
(Read "$1,200,000" as *one million two hundred thousand dollars*.)
Last year, the company invested $1.2 million in new building projects. (In large numbers, you should say *point* for the decimal. Read "$1.2 million" as *one point two million dollars*.)

PRACTICE

Match the money amounts written in words to the same amounts written in numbers and symbols. Write the letter of the correct answer on the line. Then read each amount aloud.

______	**1.** forty-five cents	**a.** $14
______	**2.** three million dollars	**b.** $566.12
______	**3.** fourteen dollars	**c.** 15¢
______	**4.** one dollar and ninety-nine cents	**d.** $3,000,000
______	**5.** fifteen cents	**e.** $75.80
______	**6.** five hundred sixty-six dollars and twelve cents	**f.** $1.99
______	**7.** seventy-five dollars and eighty cents	**g.** 45¢

Read each money amount aloud. Then write the money amount in numbers and symbols.

8. one thousand one hundred dollars ______________

9. fifty cents ______________

10. seven dollars and sixteen cents ______________

11. seven million dollars ______________

12. eighty-six cents ______________

13. three hundred thirty-seven dollars and twenty-three cents ______________

LESSON 2: Time

It is important to know how to read time correctly. We use time to tell when something begins and ends. We use time to plan our day. To know what time it is, you have to read time on a clock. Clocks show time and help us keep track of our day.

People use two kinds of clocks: digital clocks and analog clocks.

Digital Clocks

Digital clocks show time in electronic numbers. Digital clocks can be alarm clocks, timers, or even cellphone applications.

When reading a digital clock, read it from left to right. Notice that the hour comes *before* the colon (:). The minutes come *after* the colon. This time is read as "one AM."

You may also see the letters *AM* (or a.m.) or *PM* (or p.m.) on the clock. *AM* is an abbreviation for a Latin phrase that means "before midday." Hours from midnight (12:00) to 11:59 (one minute before noon) are AM hours. *PM* is an abbreviation for a Latin phrase that means "after midday." At noon, the hours become PM. Hours from noon (12:00) to 11:59 (one minute before midnight) are PM hours.

Analog Clocks

Analog clocks display time by the placement of a long hand and a short hand that point to positions around the clock. An analog clock may or may not have numbers. It may take some practice to read analog clocks.

An analog clock can be read by looking at the position of the long and short hands. The **face** is the area where the numbers 1 through 12 are spread out equally.

Follow these steps to read the time on a clock with hands.

- Look at the hour hand. The hour hand is the shorter hand. It tells you the hour.
- Look at the minute hand. The minute hand is the longer hand. It tells you the minute. Each mark on the clock face stands for 1 minute. Each number stands for 5 minutes.

12 = :00	1 = :05	2 = :10	3 = :15	4 = :20	5 = :25
6 = :30	7 = :35	8 = :40	9 = :45	10 = :50	11 = :55

Examples:

The hour hand is on 8. The minute hand is on 12. The 12 stands for :00. The clocks shows 8:00 or eight o'clock.

The hour hand is just past the 6. The minute hand is on the third mark past 2. The 2 stands for 10, so the third mark is 13. The clock shows 6:13 or six thirteen.

You see times written in words or numbers. You will often see the phrase *o'clock*. *O'clock* is a short version of the phrase *of the clock*. You will also see the phrases *half past, quarter till,* and *quarter past*.

An hour can be divided into halves. A full hour is 60 minutes. A half hour is 30 minutes. When the minute hand of a clock is on 6, the time can be written as a *half past* the hour.

Example:

You may see this time written as *2:30, two thirty,* or *half past two.*

An hour can also be divided into quarters. Quarter-hours are 15 minutes. When the minute hand is on 3, the time can be written as a *quarter past* the hour. When the minute hand is on 9, the time can be written as a *quarter to* the next hour.

Examples:

You may see this time written as *2:15, two fifteen,* or *quarter past two.*

You may see this time written as *2:45, two forty-five,* or *quarter to three*

PRACTICE

Match the clock face with the correct time.

a. 10:10	**c.** twelve o'clock	**e.** 2:30
b. quarter past three	**d.** quarter to two	**f.** seven thirty-two

1.

______

2.

______

3.

______

4.

______

5.

______

6.

______

Read the clock. Write the time on the line.

7.

____ : ____

8.

____ : ____

9.

____ : ____

10.

____ : ____

LESSON 3: Dates

Dates tell when things happened or will happen. A date may have one or more of these pieces of information.

Year	The United States celebrated its 200th year in **1976**.
Month	Independence Day happens in **July**.
Day of the month	On July **4**, people celebrate with parades and fireworks.
Day of the week	Last year, Independence Day happened on **Saturday**.

The names of months and days of the week are always capitalized.

Month (Mo.)	Abbreviation	Month Number
January	Jan.	1
February	Feb.	2
March	Mar.	3
April	Apr.	4
May	May	5
June	Jun.	6
July	Jul.	7
August	Aug.	8
September	Sept.	9
October	Oct.	10
November	Nov.	11
December	Dec.	12

You may see the name of the month abbreviated. You might also see it as a number. There are twelve months in a year. January is month 1. December is month 12.

Day	Abbreviation
Sunday	Sun.
Monday	Mon.
Tuesday	Tu., Tue., or Tues.
Wednesday	Wed.
Thursday	Thurs.
Friday	Fri.
Saturday	Sat.

You may see the name of the day abbreviated.

You see dates written with number and words.

Examples:

Tuesday, September 15, 2015
day of week → Tuesday; month → September; day → 15; year → 2015

Commas are used to separate parts of a date. Notice the comma after the day of the week (*Tuesday*). There is also a comma after the day of the month (15).

May 4, 1994
month → May; day → 4; year → 1994

Notice the comma between the day of the month (4) and the year.

1-12-2022
month → 1; day → 12; year → 2022

The month, day, and year have dashes (-) or slashes (/) between them.

A calendar is a chart that helps you keep track of dates. It allows you to see the whole month at one time. You can search for a date on a calendar quickly. A calendar can also help you to keep track of important events in your life.

Example:

JANUARY ← name of month

Sunday	Monday	Tuesday	Wednesday	Thursday	Friday	Saturday
1	2	3	4	5	6	7
8	9	10	11	12	13	14
15	16	17	18	19	20	21
22	23	24	25	26	27	28
29	30	31				

day of the week

day of month

This calendar is arranged in weeks. A week begins with Sunday and ends with Saturday. Some blocks in the calendar are blank. Since the month ends on Tuesday, the empty blocks on the last row are the first four days of February.

PRACTICE

Draw lines to match the underlined dates in word form with the correct number form.

1.	My parents arrived in Ohio on <u>June 21, 2004</u>.	**a.**	1-8-2023
2.	Shannon will turn forty on <u>January 8, 2023</u>.	**b.**	5-8-1983
3.	That picture was taken on <u>February 16, 1993</u>.	**c.**	2-16-1993
4.	My parents were married on <u>Dec. 28, 1996</u>.	**d.**	6-21-2004
5.	The poster is from a concert that took place on <u>Sept. 4, 2016</u>.	**e.**	12-28-1996
6.	Was Luis born on <u>May 8, 1983</u>?	**f.**	9-4-2016

Write the full name of the parts of the date.

7. Tues., Oct. 15 ________________, ________________ 15

8. Sat., Nov. 9, 2019 ________________, ________________ 9, 2019

9. Wed., Jul. 13 ________________, ________________ 13

10. Mon., Mar. 30 ________________, ________________ 30

11. Thurs., Apr. 1 ________________, ________________ 1

12. Sun., Aug. 21 ________________, ________________ 21

LESSON 4: Handwriting

Usually, you read printed type—in books, magazines, ads, and online. Letters in printed type are easy to read. Sometimes you read handwritten words. Each person's handwriting is slightly different, and you may have to learn to read some of the letters.

Formal handwriting is called cursive. Some cursive letters look very much like the printed type versions you usually see. The cursive letters are more rounded, but you can recognize their shapes.

Cursive	Printed Type	Cursive	Printed Type
A, a	A, a	N, n	N, n
B, b	B, b	O, o	O, o
C, c	C, c	P, p	P, p
D, d	D, d	Q, q	Q, q
E, e	E, e	R, r	R, r
F, f	F, f	S, s	S, s
G, g	G, g	T, t	T, t
H, h	H, h	U, u	U, u
I, i	I, i	V, v	V, v
J, j	J, j	W, w	W, w
K, k	K, k	X, x	X, x
L, l	L, l	Y, y	Y, y
M, m	M, m	Z, z	Z, z

Other cursive letters are shaped differently than the printed type versions. You must learn to recognize these letters. Lower-case *g*, *j*, and *q* are similar to their printed versions. But the other letter shapes are quite different.

Cursive	Printed Type	Cursive	Printed Type
F, f	F, f	Q, q	Q, q
G, g	G, g	S, s	S, s
I, i	I, i	Z, z	Z, z
J, j	J, j		

Many people use cursive on special occasions. Look for it in

- wedding invitations
- graduation announcements
- certificates and diplomas
- awards

PRACTICE

Draw lines to match the cursive letter to its printed type version. Look at the chart above if you need help.

1.	*F*	Q
2.	*z*	q
3.	*G*	S
4.	*s*	Z
5.	*q*	F
6.	*I*	G
7.	*Q*	z
8.	*S*	J
9.	*Z*	I
10.	*J*	s

Read the invitation. Then print each cursive word.

11. *Quincy* ______________________

12. *Join* ______________________

13. *Friday* ______________________

14. *sparkling* ______________________

15. *Zimmerman* ______________________

LESSON 5: Forms

A form has blank spaces where you must fill in information. Many places will ask you to fill out forms. For example, employers will ask you to fill out a form when you apply for a job. Banks will ask you to fill out a form when you apply for a loan or open an account. Shops will ask you to fill out a form when you order something that needs to be delivered.

You will often fill out paper forms, but you will also fill out forms online by typing. Follow these steps any time you fill out a form.

- Read the entire form before you start to fill it out. Pay attention to the **formatting**. Formatting may show how the form is organized. Understanding formatting can make filling out the form easier.
- Determine the purpose of the form. Each form has a specific **purpose**, or reason, for its use. You will more easily understand what kind of information is needed if you understand the purpose of the form.
- Fill out the form correctly and completely. Read the directions first. You may leave some spaces blank. You should not fill in sections that say "For office use only," "For property owner use only," or "To be completed by supervisor."
- If you cannot answer a question on a form, write *N/A. N/A* means "not applicable." This means this question does not apply to you. Do not leave lines blank.
- If you need more information in order to fill out a form, ask questions.

If you fill out a form incorrectly, you may find that processes slow down. For example, if you write your address incorrectly on a magazine subscription form, your magazine cannot reach your mailbox. If you do not include all of the requested information, the form may not be accepted. You will have to fill it out again.

Forms sometimes use abbreviations to save space. Here are some abbreviations you may see on a form.

DOB	Date of birth
SSN	Social Security Number
No.	Number
#	Number
Apt.	Apartment
Tel.	Telephone
MM/DD/YYYY	Month/Day/Year—Use two numbers for the month, two numbers for the day, and four numbers for the year. Separate the parts of the date with slashes. December 05, 2016 would be written as 12/05/2016.

Example:

This form is an **invoice**, or bill of sale. Invoices help businesses keep track of sales and customer information. Businesses use this kind of form to make sure they deliver the right kind and number of items to the customer.

Customer Order			Date	
Name				
Address			Order Number	
City, State, Zip				
Sold By	Cash	Check	Charge	Acct No
Quantity	Description		Unit Price	Amount
Received by			Tax	
			Total	

PRACTICE

Study this form. Then answer the questions.

Student Information Sheet	
Welcome to Mrs. Hu's second grade class! Please fill out this form to help me get to know your child.	
Student Information	
First Name	Last Name
What name does your child preferred to be called?	
Birthday **If your child's birthday is in the summer, would you like the class to celebrate his/her half birthday?* Yes ☐ No ☐	
Does your child have any diet restrictions or food allergies? Please list them.	
What do you think is your child's greatest strength as a student?	
What do you hope your child will learn this year?	

1. What is the purpose of the form?

____

2. If your child is allergic to milk products, where would you write that information?

____

3. If your child has a nickname, where would you write that information?

____

A common kind of form is the **subscription form**. You might subscribe to a magazine or sign up to have fresh fruit delivered to your home. Any time you fill out a subscription form, follow these steps.

- Read the entire form. Find out what you are signing up for and what information the form requires.
- Read the terms. The terms tell how much money you are required to pay and what you will receive.
- Fill in your name and address correctly so that you can receive what you buy.

PRACTICE

Study this subscription form. Then answer the questions.

Special Subscription Offer

Do you like to keep your feet warm in the winter? Do you want comfortable and stylish socks the rest of the year? Then we have a deal for you!

Our premium socks feature a new design every month. You'll enjoy designs that are colorful, bold, and surprising.

Subscribe today!

Join the Sock-of-the-Month Club!

Check one:

☐ ***Yes!*** *Sign me up for six months of socks for only $39.99.*

☐ ***Yes!*** *I want to save even more with a twelve-month subscription for only $59.99!*

Name ______________________
Address ____________________
City _______________________
State __________ Zip __________
Email ______________________

Send me

☐ Women's collection
☐ Men's collection
Shoe size __________

Payment Options

☐ Bill me ☐ Payment enclosed

Allow six to eight weeks for first month's delivery.

4. The form gives several reasons a person might want to subscribe. Tell one of the reasons.

____

5. Why is it important to fill in the line in the "Send me . . ." section?

____

6. If a person mails the card in the last week of January, in what month will she receive the first pair of socks?

____

7. Which detail on the form does the illustration of the sock support?

____

8. Complete the form. Use your information or make up information.

LESSON 6: Charts and Tables

Charts and **tables** show information clearly and briefly. They help people organize many ideas easily.

Pie Chart

A **pie chart** shows information with slices. A pie chart gets its name from its shape. It is round, like a pie. Each "slice" of the pie chart represents a percentage. All the slices add up to 100 percent.

You can quickly see which slice is largest and how the other slices compare. Right away, you know something about how the groups compare.

Some pie charts may have a label on each slice. Others may use a key to show what each slice stands for. Some pie charts will show the percentage of each slice. If the pie chart does not show the percentage, you have to estimate the percentage based on the size of the slice.

Example:

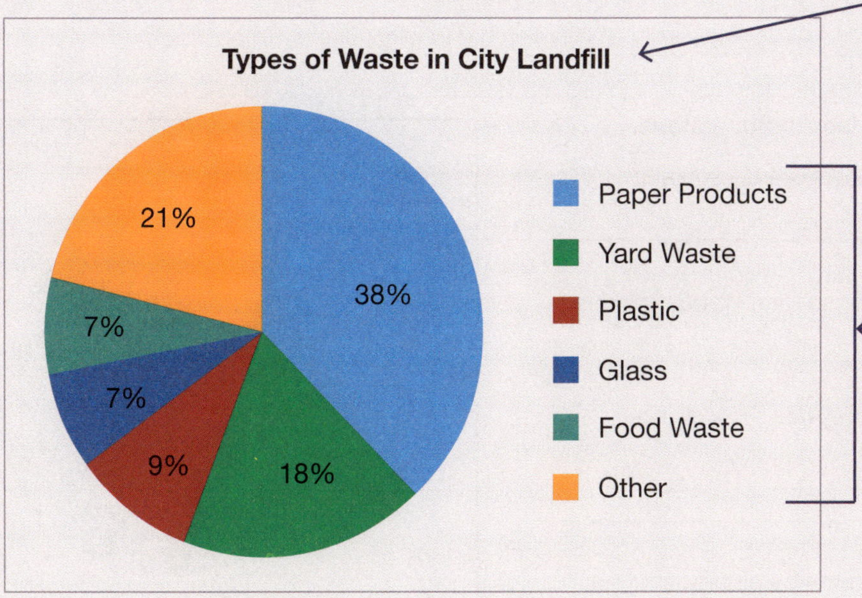

The title of a chart tells you what the pie chart is about.

This pie chart uses a key to show what each slice stands for. Each color stands for a different type of waste. For example, the green slice stands for Yard Waste. Using the key, readers know that 18% of the waste in the city landfill is Yard Waste.

Table

Tables are arranged in columns and rows. Columns run top to bottom, and rows run left to right. Each column has a heading that tells the kind of information in that column.

Example:

Tablet Model Number	Screen Size	Weight	WiFi	Camera
BX-20	5 inch	15 oz	yes	yes
CX-22	7 inch	21 oz	yes	yes
DX-30	9 inch	20 oz	yes	no
EX-50	10 inch	22 oz	yes	yes
FX-67	9 inch	17 oz	yes	yes

This table compares the features of different tablet models. Each column is labeled. To find information on a specific model, find the model number in the first column. Then read across the row to learn which features that tablet has. For example, DX-30 has a 9-inch screen. It weighs 20 oz. It has WiFi. It does not have a camera.

PRACTICE

Study this pie chart. Then answer the questions.

An Average Wedding Budget

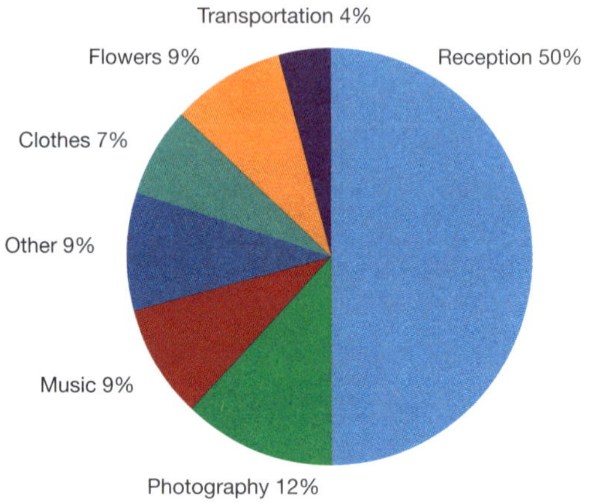

1. What is the largest expense for a wedding?

   ____

2. What is the smallest expense for a wedding?

   ____

3. What percentage of their budget do people generally spend on a photographer?

   ____

4. What three things take up the same percentage of the budget, on average?

   ________ ________ ________

5. Add up the percentages of the wedges. What is the total? ______________

Study this table. Then answer the questions.

Cheryl decided to track how many commercials she saw during her favorite evening television show. She made a table to find out what kinds of commercials she saw and how long they lasted.

Commercials Viewed During One Hour		
Subject	**How Many**	**How many minutes, combined**
food and beverage	6	3 minutes
cars	8	5 minutes
movie trailers	2	2 minutes
stores	3	2 minutes
other TV shows	4	3 minutes

6. How much of Cheryl's viewing time (one hour) was taken up by commercials?

____

7. What kind of commercials did Cheryl see most often?

____

8. What kind of commercials did she see least often?

____

9. There were six commercials for food and beverage. They lasted, all together, three minutes. About how long was each commercial?

____

10. Cheryl realized that she forgot to add one kind of commercial she saw. It was a commercial asking people to vote for a person running for mayor. It lasted 2 minutes. Add a new row to the chart to track this commercial.

Subject	How Many	How many minutes, combined

LESSON 7: Maps

A **map** is a picture of a place. This picture helps readers understand what they read. A map shows things that can be found in that place.

Maps can show very large areas. A map of a country, for example, shows states, cities, highways, and bodies of water. Maps can also show small areas. The places and sites shown on a map are in the same position as they are in the real world. By finding a place on a map, you can tell how to get to other places and sites.

The title of a map tells you the kind of information you can learn from the map. Some maps may also have a compass rose. The **compass rose** helps you find directions on a map. North (N), south (S), east (E), and west (W) are all marked on the compass rose. Some maps may also have a **map key**, or legend. A map key explains special symbols used on the map.

Examples:

This is a map of the United States. It shows state borders, state capitals, and major cities. It also shows rivers and other bodies of water.

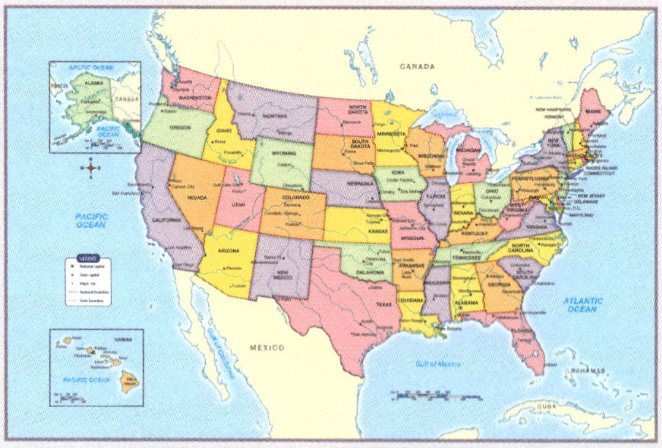

This is a map of Hermann Park in Houston, Texas. It shows popular tourist attractions. If you want to visit Hermann Park, you can read the map to see where to spend your time.

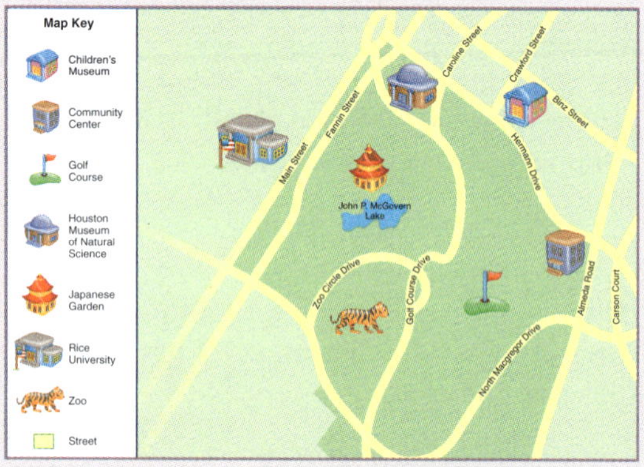

Maps can also provide other kinds of information. A map of a battleground can show where soldiers fought. A map of a house's floor plan can show how large each room is and how rooms are connected. A map of a continent can show where large groups of people migrated over time.

Example:

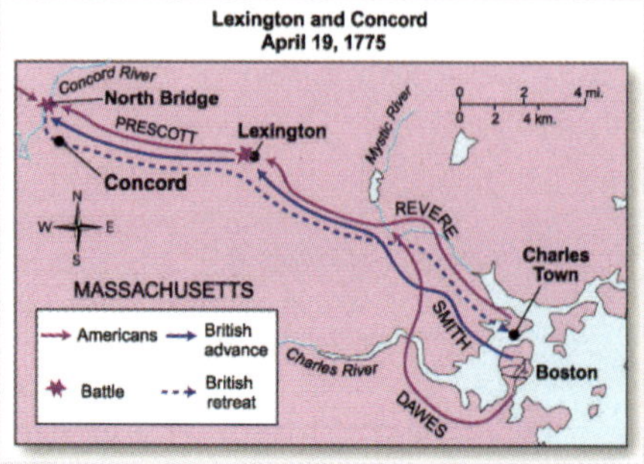

This map gives details about the Battle of Lexington and Concord. It tells about a famous battle in United States history. You can see how the battle played out by reading the map and following the arrows from town to town.

PRACTICE

Study the map. Then answer the questions.

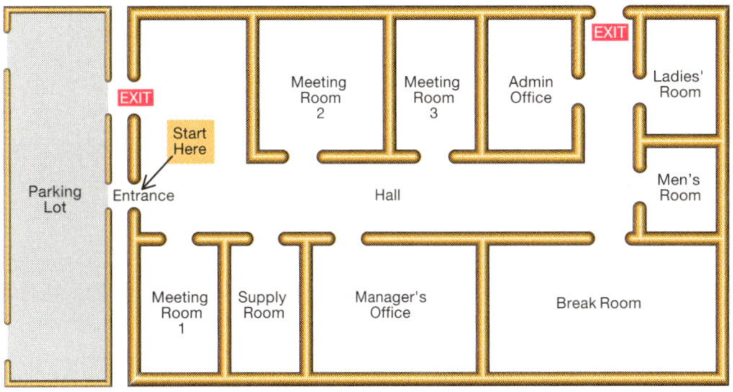

1. What area does this map show?

   ____

2. Describe the path you would take to get from the entrance to the restrooms. Use direction words like *straight, left*, and *right*.

   ____

   ____

3. How many offices does the map show? ____________
4. How many meeting rooms are available? ____________
5. Someone in Meeting Room 2 has asked you for directions to the break room. Give directions.

   ____

6. People are in the break room and need to exit the building quickly. Which exit should they use? Why?

   ____

Study the map and the map key. Then answer the questions.

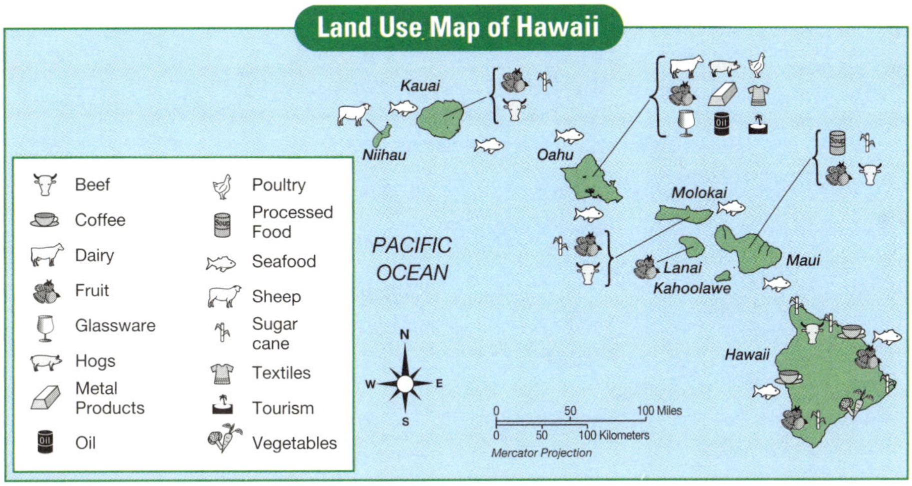

7. On which islands is land used to produce just three things? What do those islands produce?

   ____

8. Which five islands use some land for the fishing industry?

   __________ __________ __________ __________ __________

9. Where is sugar cane grown? ______________________________
10. Which island uses the most land for sugar cane planting? ____________
11. Only one island sets land aside for making metal products. Which island is it? ____________
12. What is produced on tiny Niihau? Why do you think other things are not produced on this island?

    ____

    ____

LESSON 8: Diagrams

A **diagram** is a drawing that helps people understand how a thing or process works. This drawing can help readers better understand what they read. Diagrams simplify complex ideas. Diagrams can have many forms. Here are just a few examples.

- a Venn diagram that compares different health care plans
- a drawing of bicycle gears with labels for each part
- a cut-away drawing of a car with labels for the electrical system
- instructions in images and words that shows how to put shelves together

Use these steps to read a diagram.

- Read the title of the diagram. This will tell you what the diagram is about.
- Look carefully at the images.
- Read the text and labels.
- Ask yourself, "What is the main idea, or main message, that the text and image show?"

Study this **Venn diagram**. Venn diagrams have two or more circles or ovals. They compare two or more things. A Venn diagram shows what is the same and what is different about each thing.

Example:

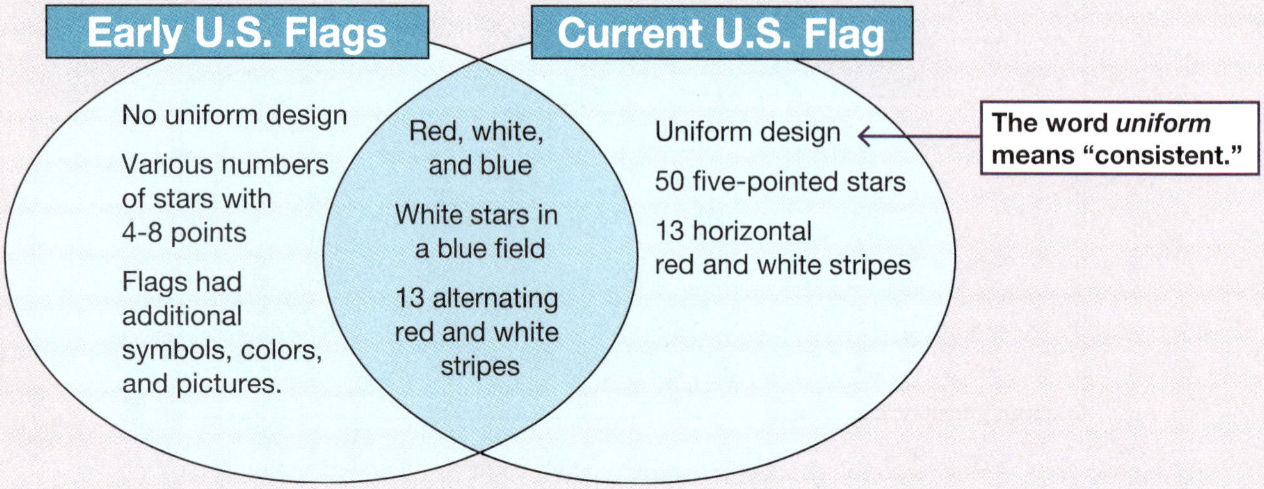

The left circle has facts about early United States flags. The right circle has facts about today's flag. The section that overlaps shows facts that are true about both earlier flags and today's flags.

PRACTICE

Use the Venn diagram on the previous page to answer the questions.

1. What traits did early flags have in common with today's flag? How do you know?

 ____

 ____

2. Early flags had fewer stars than today's flag, which has 50 stars. Think about what you know about United States history. Why did early flags have fewer stars?

 ____

Compare the facts in the left and right ovals of the Venn diagram. Then list three ways that early flags were very different from today's flags. Use complete sentences in your answers.

3. ____

4. ____

5. ____

Now look at a different kind of diagram. This diagram is a drawing with labels. This diagram shows how light enters the eye. It has labels that show the parts of an eye.

- The retina is a layer at the back of eye.
- The lens is a disk at the front of the eye.
- The pupil is a small opening in the center of the lens.
- The iris is the colored section of the eye.
- The cornea is the clear layer that protects the eye.
- The optic nerve sends messages from the eye to the brain.

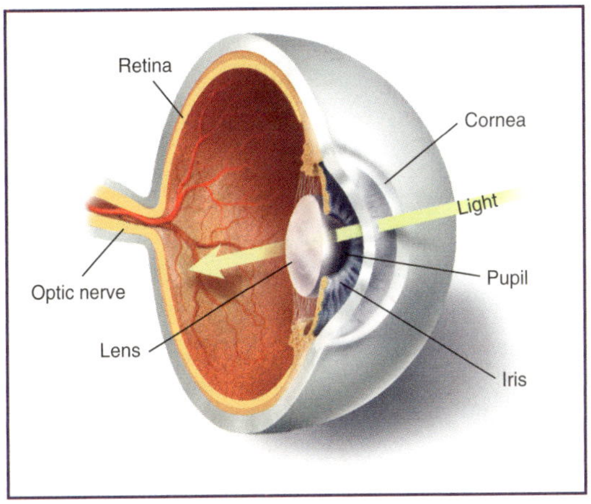

PRACTICE

Answer the questions about the diagram.

6. List the six parts of the eye that are labeled.

 ____

7. Light enters the eye through the ____________________.

 optic nerve   pupil   retina

8. What is the name of the white part of the eye? ____________________

LESSON 9: Graphs

A **graph** is a picture of information. This picture helps readers to better understand what they read. Graphs use lines and bars to show facts, numbers, or other information.

Bar Graph

One kind of graph is a **bar graph**. Bar graphs use thick lines to show information. The bars may go across the graph left to right. Or they may go up and down the graph. A bar's length matches a number or value on the graph.

To understand a bar graph, read the title and labels. Then compare the length or height of the bars. Use your finger to trace the end of a bar to the number that matches it on the graph. If the bars go across the graph, trace the end straight down to the number. If the bars go up and down the graph, trace the end straight across to the number.

Study this bar graph.

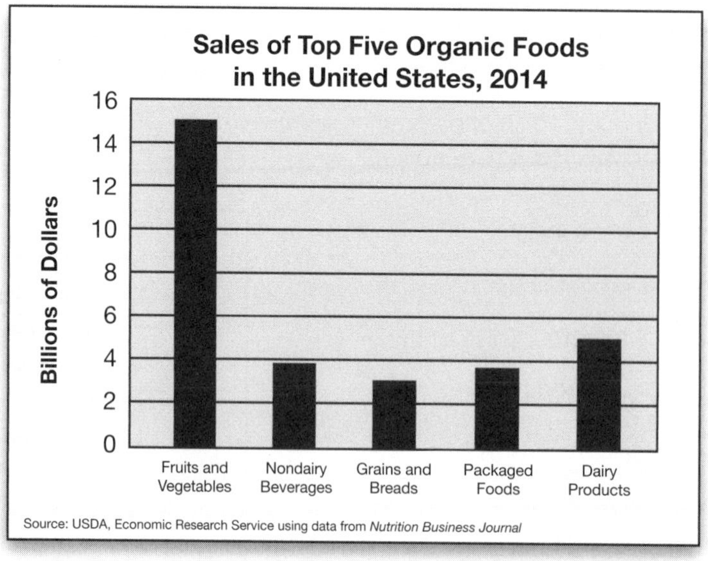

Look at the label on the left edge of the graph. It reads "Billions of Dollars." This label tells you to add "billion dollars" to each number on the left as you read. So, for 15, you would read "15 billion dollars."

Why doesn't the graph simply use the actual numbers? Why not use *15,000,000,000*? Large numbers like these are hard to read. They take up a lot of space on the graph. Graphs work to make large numbers easier to understand. So they often use shorter forms of large numbers.

Now use your finger to trace the bar for *Dairy Products* from the bottom up. Notice that it stops above the line that shows 4 billion. You must estimate the number. The bar for Dairy Products is between the 4 and the 6, so you can estimate that its value is 5. So you can say that about 5 billion dollars' worth of this kind of organic food was sold in 2014.

PRACTICE

Use the graph on the previous page to answer the questions.

1. Which food had the highest sales in 2014? ______________________
2. Which foods had nearly 4 billion dollars in sales? ______________________
3. Which food brought in the least profits in 2014? ______________________
4. Read the source information at the very bottom of the graph. How is this information helpful to readers?

____

Another kind of graph is a **line graph**. A line graph uses lines instead of bars to show numbers or values. Line graphs often show changes that happen over a period of time. Dates that show how much time is recorded appear on one edge of the chart.

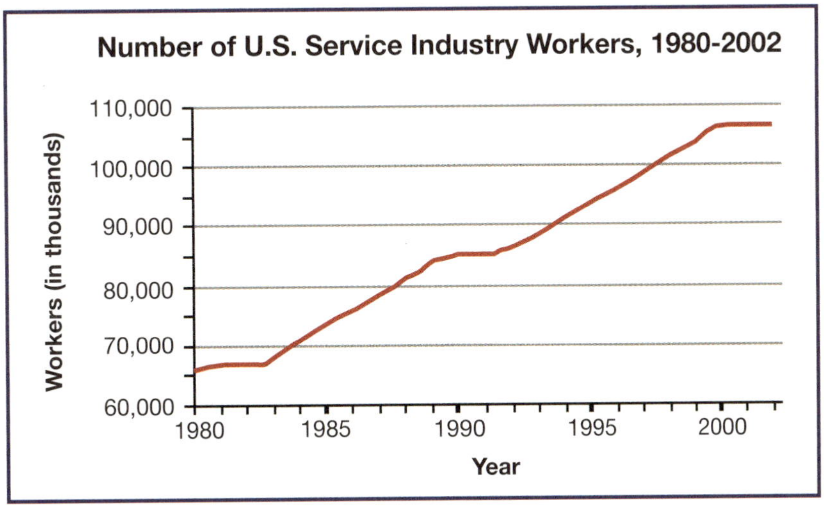

The line graph shows numbers of service industry workers from 1980 to 2002. The dates are across the bottom of the graph. Each small mark is a single year. But not all the dates will fit on the line. Instead, a number is printed for every five years. To read a small mark, count the number of marks from the last date. Then add that number of marks to that date. Two small marks past the year 1995 is 1995 + 2, or 1997.

The red line shows the change in numbers of workers. You can see that from 1980 to 2002, the number of people working in service industries goes up.

PRACTICE

Answer the questions about the line graph.

5. Look at the numbers on the left edge of the graph. What do you notice about the number of workers in the earlier years?

____

6. Look at the red line. Describe the change in the number of workers from 1980 to 2002.

____

Circle the correct answer.

1. Read the amount of money.

   two thousand fourteen dollars and two cents

   Which is the correct way to write this amount in numbers and symbols?

   A. $2,414.02

   B. $2,214.02

   C. $2,014.02

   D. $2,004.02

2. Which is true of an analog clock face?

   A. A time is marked as a.m. or p.m.

   B. The short hand points to the hour.

   C. A colon is used between the hour and the minutes.

   D. The numbers around the face are numbers of minutes.

3. Which is a correct way to write the time the clock shows?

   A. noon

   B. 12:30

   C. twelve o'clock

   D. half past eleven

4. Which date is the same as 2-12-2020?

   A. February 12, 2020

   B. March 12, 2020

   C. January 12, 2020

   D. December 12, 2020

5. Which date is circled on the calendar?

   A. Sun., Feb. 15

   B. Mon., Feb. 15

   C. Tues., Feb. 15

   D. Thur., Feb. 15

Read the sentence from a wedding invitation. Then circle the correct answer.

Mr. and Mrs. Javier Santiago
request the honor of your presence
at the wedding of their daughter,
Gabriella.

6. What is the first letter in the word *Santiago*?

   A. F

   B. G

   C. J

   D. S

7. What is the first letter in the word *Gabriella*?

   A. G

   B. I

   C. S

   D. Z

Study the form. Then circle the correct answer.

Flu Shot Form		
Personal Information		
Name: ______________________		
F: ☐ M: ☐ Date of Birth (MM/DD/YYYY): ______ *Age: ______		
Payment Method		
☐ Cash ☐ Check ☐ Credit ☐ Insurance		
****Screening Questions**		
Check one answer for each question.		
1. Are you feeling well today?	☐ Y	☐ N
2. Have you had a flu shot before?	☐ Y	☐ N
3. Have you ever had a reaction to a flu shot?	☐ Y	☐ N
4. Are you allergic to eggs?	☐ Y	☐ N
5. Can you stay in the waiting area for ten minutes after receiving the flu shot?	☐ Y	☐ N
Consent		
Signature: ______________________		
Date: ______________		
**Must be 18 years or older, unless a parent or guardian is present. **Based on your answers, the nurse may ask you more questions to determine if you can safely receive a flu shot today. If the nurse says you cannot get a flu shot, you will get a full refund.*		

8. What is the main reason people should fill out this form?

A. to check whether they are in good health

B. to show that they have paid for a flu shot

C. to find out whether they are allergic to eggs

D. to make sure they can safely get a flu shot

9. If your birthday is on January 8, 1998, how would you fill in the Date of Birth section?

A. 1/8/98

B. 1/8/1998

C. 01/08/1998

D. January 8, 1998

10. Which section of the form requires extra information depending on how a person answers the questions?

A. Personal Information

B. Payment Method

C. Screening Questions

D. Consent

11. If a person is under 18 and wants to get a flu shot, what should he or she do?

A. Fill in the number of the insurance policy.

B. Bring along a parent or guardian.

C. Pay with cash.

D. Stay in the waiting area for ten minutes.

Study the table. Then circle the correct answer.

Newspaper Sections	Types of Information
Front Page	top local news stories
Sports	events and scores
Business	financial news
Editorials	opinions
Entertainment	movies, TV, and book reviews

12. What is the purpose of this table?

A. to judge the content of the newspaper

B. to describe the sections of the newspaper

C. to persuade people to read the newspaper

D. to teach people how to write for a newspaper

13. You heard about a bank robbery in your town. Which section would you check to learn more?

A. Sports

B. Entertainment

C. Front Page

D. Editorials

Study the pie chart. It shows how federal money, such as funds collected as taxes, is spent. Then circle the correct answer.

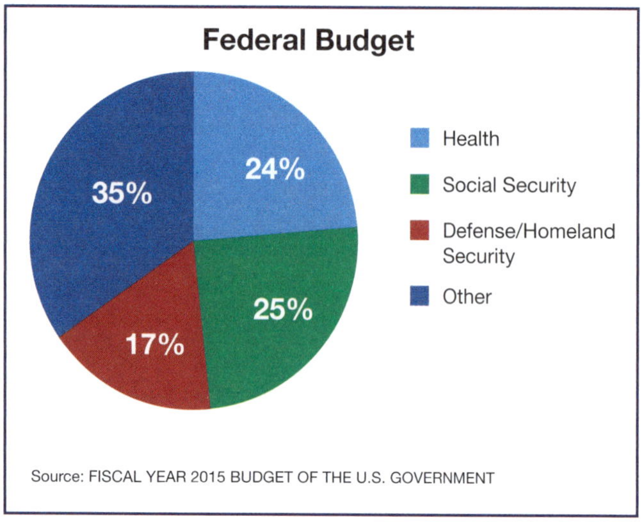

Source: FISCAL YEAR 2015 BUDGET OF THE U.S. GOVERNMENT

14. What is the largest expense in the federal budget?

A. Other

B. Health

C. Social Security

D. Defense

15. How much of the budget do Health and Social Security combined take up?

A. 42%

B. 49%

C. 59%

D. 60%

16. What percentage of the budget is spent on Defense/Homeland Security?

A. 17%

B. 24%

C. 25%

D. 35%

Study the map. Then circle the correct answer.

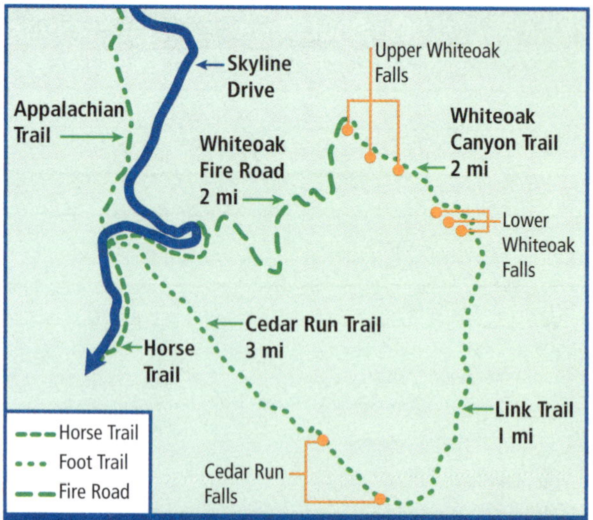

Whiteoak Canyon Area

Road and Trail Map (Shenandoah National Park, Virginia)

17. Which trail is open to people on horseback?

A. Link Trail

B. Horse Trail

C. Whiteoak Canyon Trail

D. Whiteoak Fire Road

18. Which trail connects Lower Whiteoak Falls with Cedar Run Falls?

A. Link Trail

B. Skyline Drive

C. Cedar Run Trail

D. Whiteoak Canyon Trail

19. Which trail runs alongside the left of Skyline Drive?

A. Cedar Run Trail

B. Appalachian Trail

C. Whiteoak Fire Road

D. Whiteoak Canyon Trail

Study the diagram. It explains how compost, or humus, is made. Compost is rich soil that helps plants grow. Read the diagram from the top to the bottom. Then circle the correct answer.

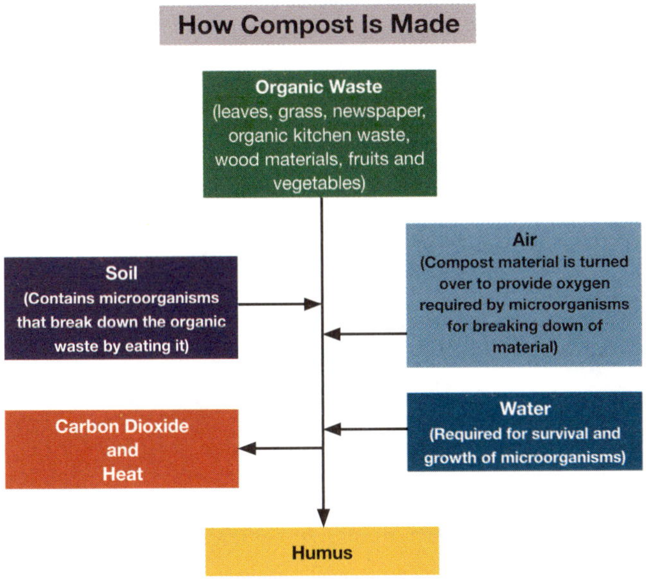

20. What do microorganisms need to break down waste into compost?

A. air

B. heat

C. leaves

D. soil

21. If you have a compost heap, what must you do to help the waste break down into compost?

A. Save old newspapers.

B. Put worms in the pile.

C. Mix the material up so air gets in.

D. Add carbon dioxide to the material.

22. Why is soil needed in the process of making compost?

A. Microorganisms live in the soil.

B. Water is in the soil.

C. The soil is warm.

D. The soil gives off air.

Study the graph. Then circle the correct answer.

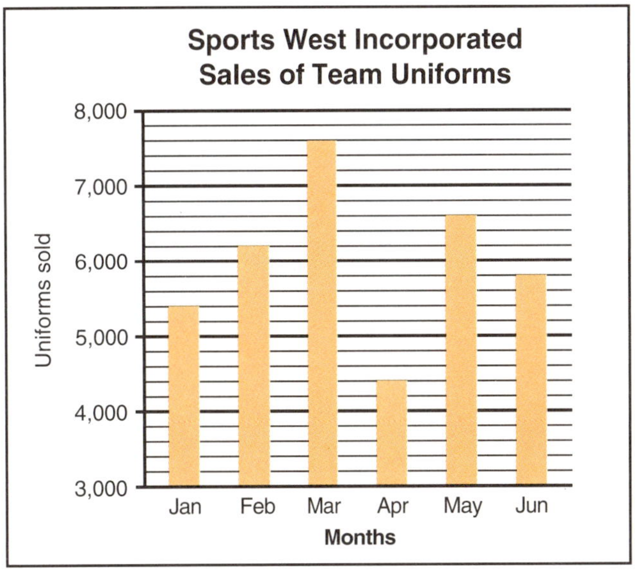

23. What kind of information is shown at the bottom of the graph?

A. sports seasons

B. a range of time

C. sales of uniforms

D. kinds of uniforms

24. In which month did the company have the most sales?

A. February

B. March

C. April

D. May

25. In which two months were sales the closest?

A. May and June

B. March and June

C. January and April

D. February and June

LESSON 10: Indexes

An **index** is a short list of subjects that are in a book. It tells where you can find information about each subject in the book.

Indexes are arranged in ABC order. They may include subjects or people's names. An index shows one or more page numbers for each entry. Check those pages for information about the topic that interests you.

If a book has different kinds of information about a subject, you may see smaller entries, or subentries, under the entry for that subject. Entries or subentries may also refer to other entries that have related information.

To use an index, turn to the back of the book. Find the subject you are looking for using ABC order. Subentries also appear in ABC order within an entry.

L
Lettuce, 116
  iceberg, 117
  Romaine, 118
  wilted, salad, 215
Lima beans ← **entry**
  Creole, 134
  green, 134
  Italian, 342
Lime
  chiffon pie, 301
  garnishes, 124
Liver, beef, 181 *(See also Chicken livers.)*
  and onions, 182
Lobster
  boiled, 264
  salad, 217
  tails, broiled, 260
London broil, 244

M
Macaroni
  and cheese, 88
  casserole, 360 ← **subentries**
  salad, 216
Macaroons, 323
Mango, 111
Manicotti, 94
Marble cake, 311
Mashed potatoes, 66
Measuring ingredients, 8–10 ← **pages in book where subject is found**
Meats *(See also specific kind.)*
  carving, 32
  cuts, 26
  freezing, 30
  storing, 31
Meat loaf, 100
Meatballs, 162
Melon, 114 *(See also Cantaloupe, Honeydew, Watermelon.)* ← **related entries**

PRACTICE

Study the index above. Then answer the questions.

1. On which page would you likely find instructions on how to mash potatoes smoothly?

____

2. If the indexer (the person who created the index) added an entry for limeade, what entry would it follow?

____

Read each sentence and circle the correct phrase or number to complete the sentence.

3. This part of the index does not have page numbers for _________.
   how to measure ingredients   how to freeze meats   how to cook chicken livers
4. You would likely find a recipe for a meat dish on page _________.
   31   100   134

LESSON 11: Table of Contents

Most books have a **table of contents**. A table of contents is at the beginning of a book. It shows the order of the parts of a book or magazine. It lists the page numbers on which each part begins. The table of contents of a book usually lists chapters or sections.

A book's table of contents is a good place to start when you read. The table of contents gives you a road map. It tells you what you will read about as you go through the book.

In nonfiction books, the chapter titles often hint at what each chapter is about. But in some books, such as novels, chapter titles may simply be "Chapter 23" or "23." In that case, the table of contents is useful mainly to find the page on which a chapter begins.

PRACTICE

Study the table of contents from a book on choosing and training a dog. Then answer the questions.

Table of Contents

1. How many chapters of the book discuss things to think about before readers get a dog? How do you know?

____

2. In which chapter would you likely find information about if it is a good idea to adopt a rescue pet?

____

3. To which page should you turn if you want to learn about supplies a new puppy or dog needs? Which page might help you purchase them?

____

4. Which section of the book likely speaks with praise about the relationship that dogs and people can develop?

____

5. Which chapter might contain a form to use to record a dog's vaccinations?

____

6. Which chapter's title suggests that you should think carefully about the size of dog you need?

____

7. Which section of the book warns readers to consider whether they really want a dog?

____

The table of contents may include titles of section within a chapter. You can use these section titles to find specific information quickly. This is important in a book you may turn to frequently to review information that you use often.

Look at the section titles for one of the chapters in the book on dogs. Then answer the questions.

8. If your dog often barks loudly at people and other dogs, which section would likely tell you how to fix the problem?

____

9. Which section might include advice on cleaning up after your dog and other questions about taking your dog out in public?

____

10. Before you decide which section on leash training to read, what must you know?

____

LESSON 12: Ordered Lists

Lists are helpful and popular tools. You probably use lists often to keep track of things you need to do, items you need to buy, and events you plan to attend.

One kind of list is the **ordered list**. An ordered list presents useful information in a brief, clear way. Some are in ABC order. Others are in number order.

Look at the delivery work schedule for Speedy Fast Package Delivery company. It lists the names of the drivers in ABC order by last name. It also lists the route that the driver has been assigned to work for that week.

Example:

Speedy Fast Package Delivery Employee Weekly Work Schedule	
Employee Name	**Assigned Route**
Andrus, Jane	Route 5
Conway, Leticia	Route 1
Erickson, Martin	Route 6
Le Herrera, Sarah	Route 3
Reyes, Omar	Route 2
Zanzibar, Gerrard	Route 4

Pretend you are Martin Erickson, an employee at Speedy Fast Package Delivery company. You need to know which package delivery route you will be working this week. You do not have to read all of the names on the schedule to find out which route you work. Your last name begins with an *E*. You look for Erickson, Martin. It is between Conway, Leticia and Le Herrera, Sarah. Once you find your name, you look across to see that you are delivering on Route 6 this week.

Another kind of list is a directory. A directory lists phone numbers or locations of offices within a building. Office directories are helpful in buildings that have a lot of businesses on many floors. If you know which office you need, you can easily find your way.

Example:

The Healing Arts Building Directory			
Building 100-A		**Building 200-A**	
Suite 101-A	Dr. Lori Smith, M.D.	Suite 202-A	Dr. Lin Tam, M.D.
Suite 103-A	Dr. Ali Shazah, M.D.	Suite 204-A	Dr. Nam Pham, M.D.
Suite 105-A	Dr. Miguel García, M.D.	Suite 206-A	Dr. Julio Katz, M.D.
Suite 107-A	Dr. Veronica Haag, M.D.	Suite 208-A	Dr. Lynn Baers, M.D.

Imagine you are visiting your new doctor for the first time. You know that your new doctor is located in Suite 103-A, but you do not remember the doctor's name. You can use the directory to find the doctor's name. Dr. Ali Shazah, M.D. is in Suite 103-A.

PRACTICE

Write the names of the cars in the box in ABC order.

Toyota	Jaguar	Acura	Honda	Lexus	Audi
Kia	BMW	Mercedes	Cadillac	Ford	Infiniti

A	D-H	K-L
1. ____________	**5.** ____________	**9.** ____________
2. ____________	**6.** ____________	**10.** ____________
B-C	**I-J**	**M-Z**
3. ____________	**7.** ____________	**11.** ____________
4. ____________	**8.** ____________	**12.** ____________

Read the lists of ordered times of marathon races for both groups of women. Circle the name that completes each sentence. Then answer the questions.

13. For women 19-39 years old, the best time was held by ____________.

Boyd Anders Zion

14. For women 40 years old and older, the best time was held by ____________.

Knish Dresden Jesnick

15. If the qualifying time for the final race is 2 hours and 40 minutes, how many runners will be going to the final race? How many will not be going to the final race?

____

____

Women, 19-39 years old	
Time	**Name**
2:35	Anders, L.
2:37	Zion, E.
2:45	La Rouge, F.
3:00	Boyd, K.
Women, 40 years old and older	
Time	**Name**
2:50	Jesnick, I.
2:55	Hernández, M.
2:58	Knish, S.
3:20	Dresden, C.

LESSON 13: Grouped Lists

Another kind of list is the **grouped list**. These are lists in which similar things are grouped together. Each group has something important in common.

To use a grouped list, read the list's title and headings first. These tell you the reason for the list and the kinds of things you will find. Think about why some items are together under one heading (Put on ice in cooler) while other items are under another heading (Pack in backpack). The groups help you sort information and take action.

PRACTICE

Study the grouped list. It organizes items to take to a picnic. Look at the headings and the items under each heading. Then answer the questions.

Picnic Necessities

Put on ice in cooler:
- Meats
- Cheeses
- Mayo and mustard
- Pickles
- Drinks

Pack in backpack:
- Ground cloth
- Plates and forks
- Napkins
- Sunscreen
- Insect repellant
- Toys to toss and catch

Ask people to carry:
- Boom box and CDs
- Beach umbrella
- Folding chairs

1. Why do the items in the first group need to be kept "on ice"?

____

2. List two things that the items that go in the backpack have in common.

____

3. Why are the items in the "Ask people to carry" group not in the "Pack in backpack" group?

____

4. To which group would you add slices of watermelon? Why?

____

5. To which group would you add salt and pepper? Why?

____

Some grouped lists organize items. Others organize actions. Lists that organize actions are often called **to-do lists** because people need to do the tasks on them. To-do lists can be grouped in several ways. Perhaps some tasks need to be done at the same time. Perhaps they need to be done at the same place.

As with any grouped list, study the list. Determine why the tasks are grouped. What do the tasks in each group have in common?

PRACTICE

Read the to-do list. Think about why the tasks are grouped as they are. Then answer the questions.

At home:
✓ Fold laundry and take to kids' room
✓ Check pantry for foods that have gone bad, and throw them away
✓ Send bills out

In town:
✓ Meet with Lissa's teacher at noon
✓ Pick up milk and cat food from the store
✓ Pick up paperwork at doctor's office
✓ Get gas

At game:
✓ Set out snacks
✓ Monitor kids during break time

6. Why are two tasks grouped under the heading "At game"?

____

7. Why does it make sense to take care of all the "In town" tasks in the same time period?

____

____

8. To which group would you add "Pick up kids from school"? Why?

____

____

9. "Call dentist to make appointment" is another task. Could it go in more than one group? Where would you put it? Explain.

____

____

10. How else could you group tasks in a to-do list?

____

____

LESSON 14: Dictionaries

What can you do when you do not know a word? You can start by looking in a dictionary. A **dictionary** is a book that lists words in ABC order. It gives information for each word. A dictionary can do the following.

- tell how to say, or pronounce words
- tell the plural (more than one) spelling of words. The abbreviation *pl.* often comes before a plural word.
- tell what words mean
- tell how to use words
- give examples of how some words are used

Guide words at the top of each page tell you the first and last words on that page.

Each dictionary entry uses the same **format**, or arrangement. Learn this format so that you can quickly find the information you need.

Study the format of the dictionary entry for the word *complete*.

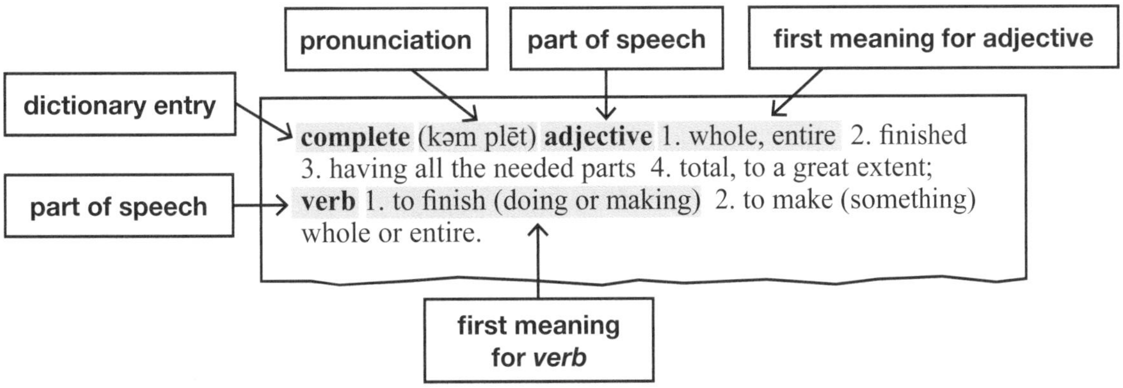

You can see that *complete* can be used as an adjective, to describe nouns, or as a verb, to show action. *Complete* has four meanings as an adjective and two as a verb.

PRACTICE

Answer these questions about the dictionary entry for *complete*.

1. Which meaning of *complete* might be used to describe a painting that is ready for display?

 ____

2. Which part of speech is *complete* in this sentence? "Tom will <u>complete</u> the report in time for tomorrow's meeting."

 ____

3. Which meaning of *complete* is used in this sentence? "The fundraiser was a <u>complete</u> success!"

 ____

4. Write a sentence using the first adjective meaning of *complete*.

 ____

____

Some dictionaries include entries for famous people, objects, places, and documents. Others do not include any proper nouns (names of particular people, places, and documents). If you need to know a little about scientist Marie Curie, for example, you might find her name, dates, and major achievements in a large dictionary. If not, check another reference work, such as an encyclopedia.

Now study several other dictionary entries.

Liberty Bell The bell in Independence Hall, Philadelphia, that was rung on July 8, 1776, to mark the signing of the Declaration of Independence.

li·brar·y |lī'brer'·ē *or* lī'brə·rē| *n., pl.* **li·brar·ies** **1** A collection of books, magazines, newspapers, and other materials, especially when arranged and catalogued for public use. **2** A building, room, or other place in which such a collection is kept.

lick |lik| **1** *v* To pass the tongue over the surface of: Be sure to *lick* each stamp. **2** *v.* To move or pass over or about: The flames *licked* the coals. **3** *n.* The act of licking: My dog gave me a *lick* on my face. **4** *n. informal* A quick or careless action or effort. **5** *n. informal* A bit or small amount: We did every *lick* of work on it ourselves. **6** *v. informal* To defeat or thrash: We *licked* them easily. **7** *n. informal* A sharp blow or hit. **8** *n.* A salt lick.

Notice that the three entries are in ABC order. Also, the part of speech is abbreviated rather than spelled out. *Noun* is shortened to *n. Verb* is shortened to *v. Plural* is shortened to *pl.* You will often see these and other abbreviations in dictionary entries. If you need help, check the front pages of the dictionaries. All the abbreviations and the words they stand for will be listed there.

PRACTICE

Answer the questions about these entries.

5. Which word has the most meanings listed? How many meanings are given for that word?

____

6. Which entry names a famous object and tells about its history?

____

7. In the entry for Liberty Bell, *liberty* is not defined. If a reader did not know the word *liberty*, what could the reader do?

____

8. Which word can mean both a building and what is kept in that building?

____

9. Which two parts of speech can the word *lick* be used as?

____

LESSON 15: Glossaries

A **glossary** is a list of important words in a book. The words are used to talk about the book's subject. For example, a cookbook's glossary includes words that name ingredients, techniques, and kitchen tools. A biology textbook's glossary includes scientific terms that name life processes, cell parts, and lab methods.

You can think of a glossary as a brief dictionary made just for the book it appears in. The glossary usually comes at the end of the book, near the index. Some textbooks may include a glossary at the end of each chapter.

Wherever you find the glossary, you will see that the words in it are in ABC order, just as they are in a dictionary. You can use the same skills to read a dictionary and a glossary.

Look at a section from a glossary. Read each entry. You see the same pattern as in a dictionary entry: the word, pronunciation, part of speech, and meanings.

GLOSSARY

arbor [är′bər] n. A place shaded by trees.
arboretum [är′bə • rē′təm] n. A plot of land on which trees are studied and/or displayed.
botany [bot′ə • nē] n. The science that collects, studies, and explains facts about plants and plant life.
cambium [kam′bē • əm] n. The soft tissue in such trees as the maple, elm, and ash, from which new wood and bark develop.
chlorophyll [klôr′ə • fil] n. The green coloring matter in plants.
coniferous [kō • nif′ər • əs] adj. Any cone-bearing tree, such as pine.
deciduous [di • sij′ o͞o• əs] adj. Any tree that sheds its leaves each year.
pith [pith] n. The loose spongy tissue inside the stem of a tree.

PRACTICE

Use the glossary section to answer the questions.

1. This glossary is found at the end of a book. Based on the entries, what do you think the book's subject is?

___

2. Which two entries define words that name parts of a tree?

___

3. Which two entries define words that name kinds of trees?

___

4. Which word names the field of science that studies trees?

____

5. Review the meaning of the word *arbor*. Then write a sentence using this noun.

____

Now study a section of a glossary from another book. The entries are words that have many meanings in the dictionary, but the glossary gives only the meanings that are used in the book. This glossary gives the part of speech, but it does not tell readers how to say the words.

This glossary follows a chapter of the book and contains words used just in that chapter. See if you can guess what the chapter is about.

attract *verb* to pull or draw something toward
electron *noun* a very small particle of matter that circles an atom's nucleus; electrons carry a negative charge
electrostatic force *noun* the force that exists between two charged electrons and protons that attracts or repels
magnet *noun* a piece of metal that is charged so that it attracts some other kinds of metal
negative charge *noun* an electrostatic force that repels other negatively charged matter and attracts positively charged matter
pole *noun* one of the ends of a magnet
positive charge *noun* an electrostatic force that repels other positively charged matter and attracts negatively charged matter
proton *noun* a very small particle of matter that is part of an atom's nucleus; protons carry a positive charge
repel *verb* to push or propel something away from

If you guessed that the chapter discusses how magnets work, you guessed right.

PRACTICE

Answer these questions about this section of the glossary.

6. Which two entries define verbs that have opposite meanings?

____

7. Which two nouns define tiny bits of matter?

____

8. How are the terms *positive charge* and *negative charge* related to the term *electrostatic force*?

____

9. How is a pole related to a magnet?

____

Use the index to answer the questions. Circle the correct answer.

Index

1. The indexer wants to add an entry for the miller's knot. Which entry should it follow?

A. after the entry for Materials

B. after the entry for Mariner's knot

C. after the entry for Jury mast knot

D. after the subentry for natural fibers

2. Which entry has subentries and is mentioned in a "See also" reference?

A. Hitches

B. Materials

C. Half hitch knots

D. Lanyard knot

3. On which page would you begin reading to learn about the history of knots?

A. 2

B. 17

C. 19

D. 3

4. Which pages likely contain important safety information?

A. 110 and 111

B. 19–24

C. 28–34

D. 81

5. Which subentry is *least likely* to have instructions about tying a kind of knot?

A. overhand

B. nylon blends

C. knife lanyard

D. two-stranded

Use the table of contents to answer the questions. Circle the correct answer.

6. What page would you turn to for information on how to pay for a child's future schooling?

A. 65

B. 73

C. 117

D. 125

7. Which section title gives information that encourages readers to think about the distant future?

A. The Short Answer: Probably Not!

B. The Quick Fix: A Weekly Budget

C. What About the Kids?

D. The Long Haul: Make a Will

8. Which pages discuss questions about spending and saving?

A. 64–101

B. 102–104

C. 105–116

D. 125 and following

9. Which section would you be sure to read if you need to know about health and life insurance?

A. Are You Saving Enough?

B. The Long Haul: Planning for Retirement

C. The Short Answer: Insure Them

D. The Quick Fix: Set Up a College Fund

10. The next chapter's title is "Question 6: Are You Ready for Tax Season?" Which section title would you likely find in this chapter?

A. Investing Basics

B. Pay Off Your House

C. Family Time: How to Save for a Vacation

D. Bring Out Your W-2s: It's Time to Get Organized

Read the grouped list and the ordered list. They go together to teach readers how to do something. Then circle the correct answer.

Host an Evening Around the Campfire!

Now that the weather is cooler, the sun sets sooner, and the stars seem to shine brighter. It is the perfect time for a campfire with friends and family. Choose a site at a nearby state or city park where fires are allowed. Gather your supplies, and get ready for a happy evening around a glowing fire!

Fire supplies:

Tinder—small twigs, pine needles, dry grass (you can gather this on site)
Several logs, cut for the fire ring, dry and seasoned
Bucket of water (or sand, if you are in the desert) to place by the fire, in case of emergency
A shovel to break up cinders and coals when the evening is done

Snack supplies:

Thermoses of hot chocolate or cider
Cups
Ingredients for s'mores—marshmallows, chocolate squares, graham crackers
Water bottles
Marshmallow toasting sticks (optional—or use slender, long branches and burn them afterward in the fire)

Comfort supplies:

Warm jackets and shoes
Warm caps
Folding chairs
Musical Instruments—guitars, ukuleles, harmonicas
Stories and jokes

In order to make your camping trip a success, be sure to assign your friends important tasks. To make it easy, each friend can be assigned a task based on his or her first name. Use the following list as a guide to assign your friends tasks during the trip.

First Name Begins with the Letter:	Task Assigned
A–D	Build the campfire
E–J	Distribute drinks and food
K–M	Set up chairs around campfire
N–P	Gather campfire materials for the fire
Q–R	Help toast s'mores
S–V	Attend to the campfire during visit
W–Z	Extinguish the campfire when finished

11. Which items on the grouped list are not things that readers can actually pack?

A. tinder and logs

B. hot chocolate and cider

C. warm jackets and hats

D. stories and jokes

12. Which friend would be assigned to build a campfire?

A. Kristin

B. Lamar

C. Ron

D. Akira

13. Which items from the grouped list suggest that cool weather is right for campfires?

A. tinder and logs

B. cups and water bottles

C. warm jackets and hats

D. guitars and harmonicas

Study the dictionary entries. Read the guide words at the top of the page. These will help you to find information quickly. Circle the correct answer.

cherry **chip**

cher·ry [cher'ē] *n., pl.* **cher·ries** 1 A small, round, eatable fruit, red, yellow, or nearly black in color, and having a single pit. 2 The tree bearing this fruit. 3 The wood of this tree. 4 A bright red color.

child [chīld] *n., pl.* **chil·dren** [chil'dren] 1 A baby. 2 A young boy or girl. 3 A son or daughter. 4 A person from a certain family.

14. Look at the guide words at the top of the page on each corner. What is the last word that appears on this page?

A. cherry

B. child

C. children

D. chip

15. Which meaning of *cherry* is used in this sentence?

Her cherry-colored scarf stood out against the snowy streets.

A. meaning 1

B. meaning 2

C. meaning 3

D. meaning 4

Study this section of a glossary. Then circle the correct answer.

accumulate *verb* to build up, to pile up; snow and ice accumulate in a glacier

advance *verb* to move down toward the valley; a glacier advances when its ice builds up more quickly than it melts

cirque *noun* a hollow area in a mountain where ice can collect to form a glacier

erode *verb* to wear away gradually; *noun* **erosion**

glacier *noun* a large mass of ice that forms on land; glaciers slowly slide downhill, pulled by their massive weight

moraine *noun* debris dragged by a glacier as it moves; moraine ranges from little pebbles to large boulders

retreat *verb* to move back toward the mountain; a glacier retreats when its ice melts more quickly than it builds up

terminus *noun* the edge of a glacier, where ice melts as quickly as more ice flows down

16. Which two terms have *opposite* meanings?

A. advance, erode

B. advance, retreat

C. advance, terminus

D. advance, accumulate

17. If a winter brings lots of snow and ice, what will happen to the glacier?

A. It will erode.

B. It will retreat.

C. It will advance.

D. It will remain still.

LESSON 16: Simple Sentences

A **sentence** is a group of words that tells a complete idea. The words in the sentence should be in an order that makes sense.

Examples:

Mac joins the softball team.
The team practices every Monday.

Sentences begin with a capital letter and end with an end mark. Many sentences end with a period (.). Some sentences end with a question mark (?) or an exclamation point (!).

Examples:

Mac joins the softball team**.**
Did Mac join the softball team**?**
Mac joins the softball team**!**

Every sentence has a **subject** and a **predicate**. The subject tells the person or thing the sentence is about. It is usually at the beginning of a sentence.

Examples:

person → Mac joins the softball team.

group → The team practices every Monday.

To find the subject of a sentence, ask, "Who or what does something?"

Mac joins the softball team.
Who does something? Mac does something.

The **predicate** tells what the subject is or does. It tells what happens. The predicate starts with the **verb**, the action that is done by the subject. It is usually the last part of the sentence.

Examples:

Mac joins the softball team. ← **what happens**

The team practices every Monday. ← **what happens**

To find the predicate of a sentence, ask, "What happens in the sentence?"

The team practices every Monday.
What does the team do? What happens? The team practices every Monday.

A **simple sentence** has one subject and one predicate. Simple sentences are often short. Here is a simple sentence. It has a single subject and a single predicate.

Example:

Mac caught a fly ball.

Mac is the subject. The predicate is *caught a fly ball*.

PRACTICE

Read each sentence. Decide whether the underlined part is the subject or predicate. Circle the correct answer.

1. The phone rang loudly. (subject, predicate)
2. Fran left the table. (subject, predicate)
3. She talked on the phone. (subject, predicate)
4. Her coffee cooled off. (subject, predicate)
5. She poured a fresh cup. (subject, predicate)

Read each sentence. Then answer the question.

6. Fran's mother called.
   Find the subject. Who called? ____________________
7. She had lunch with Fran.
   Find the predicate. What did she do? ____________________
8. Fran met her mother at the café.
   Find the subject. Who met her mother at the café? ____________________
9. They ate healthy salads.
   Find the predicate. What did they do? ____________________
10. The trees shaded the café table.
    Find the subject. What shaded the table? ____________________

Use the subject and predicate to write a simple sentence. Remember to capitalize the first letter of the sentence and to use a period at the end of the sentence.

11. subject—Jesse; predicate—walks

    ____

12. subject—The stores; predicate—open

    ____

LESSON 17: Compound Sentences

Remember that every **sentence** has a subject that tells the person or thing that the sentence is about. Every sentence has a **predicate** that tells what the subject of the sentence is or does.

A **simple sentence** has one subject and one predicate. Simple sentences are often short.

You can combine two or more simple sentences to make a **compound sentence**. Join the simple sentences with a comma and a connecting word.

Examples:

Simple sentence — Victor likes Karen.
Simple sentence — He asks her on a date.

simple sentence | **comma + connecting word** | **simple sentence**

Compound sentence — Victor likes Karen**, so** he asks her on a date.

The connecting words that join simple sentences are called **coordinating conjunctions**. Read the seven coordinating conjunctions in the table below. What does each show about how the ideas are connected?

Examples:

Coordinating Conjunction	How the Conjunction Connects Ideas in Sentences	Example
and	shows that ideas are alike in some way	I put on my gloves, and Chet put on his mittens. (The actions are alike.)
but	shows that ideas are different in some way	My gloves are red, but Chet's mittens are blue. (The colors are different.)
yet	shows that ideas are unexpectedly different	The sun was bright, yet the wind blew cold. (Sunshine usually brings warmth.)
for	shows that cause and effect connect the ideas	Matt wore a scarf, for the cold was extreme. (The cold caused Matt to wear a scarf.)
so	shows that a result or cause connects the ideas	The forecast called for more snow, so Lina hurried home. (Lina went home as a result of the forecast.)
or	shows a choice or option	Paul can take the train, or he can walk ten blocks. (Paul has a choice—to walk or to go by train.)
nor	shows a negative choice	We did not mind the walk, nor did we want to pay for a ticket. (The speaker chooses not to pay for a ticket.)

You will read and write these four conjunctions often.

Examples:

and	The doorbell rang, **and** the dog barked loudly.
but	The clouds gathered, **but** no rain fell.
so	We are out of milk, **so** I must go to the store.
or	We can play tennis, **or** we can take a walk, **or** we can stay home.

Put the comma after the last word of the first simple sentence when creating a compound sentence. Compound sentences make writing interesting and connect ideas clearly.

PRACTICE

Read each sentence. Think about how the ideas in the sentence are related. Review the chart on the previous page, and write a coordinating conjunction that connects the sentences meaningfully.

1. We will travel to California someday, ________ we will get to see the giant redwood trees.
2. The trees grow in places near the ocean, ________ they also grow in the mountains.
3. People once used redwood trees to make furniture, ________ now many trees are protected in parks.
4. One tree, The General Sherman, is one of the largest trees on earth, ________ it is not the tallest.
5. Visitors stare at the tree in awe, ________ it may be more than 2,000 years old.

Read the three simple sentences. Think about how the ideas in the sentences are related. Review the chart on the previous page. Then write two coordinating conjunctions that connect the sentences meaningfully.

6. Redwoods grow in Yosemite National Park, ________ we want to drive there, ________ it is a long way away.
7. The park can be crowded, ________ it is less crowded in winter, ________ we will probably go in January.
8. We do not want to fly, ________ do we want to take a train, ________ we will plan on two days to drive.

Read the two simple sentences. Use a comma and coordinating conjunction to combine them into a single compound sentence. Write the new sentence. Be sure to put the comma after the end of the first simple sentence and before the conjunction.

Example: The clouds gathered. No rain fell. **The clouds gathered, but no rain fell.**

9. Driving vacations can be fun for the family. They cost less than flying.

____

10. You should check your car before you leave. You might have car trouble on the road.

____

11. You want to save money. Pack water and snacks to eat in the car.

____

12. You may know the way. A GPS device or a phone app can help if you get lost.

____

LESSON 18: Adjectives That Modify Nouns

An **adjective** is a word that modifies, or describes, a noun. Adjectives can tell how many, what color, or what size or shape. They can also describe how someone or something feels, sounds, tastes, or smells.

Examples:

The fresh bread has a delicious taste.

Fresh tells what kind of bread. *Delicious* describes the bread's taste.

Many people clapped for the talented singer.

Many tells a general number. *Talented* tells what kind of singer.

Little green buds appeared on the trees when the warm weather returned.

Little tells what size of buds, and *green* tells their color. *Warm* tells what kind of weather.

Adjectives can also compare two or more things. You must change the adjective's form to make it compare things.

Add *-er* to most short adjectives to compare two nouns.

Example:

fresh + *er* The bread baked today is fresher than the bread baked yesterday..

Add *-est* to most short adjectives to compare more than two nouns.

Example:

fast + *est* The fastest runner wins the race.

Change the *y* to *i* before adding *-r* or *-est* to adjectives that end in a consonant + *-y*.

Examples:

Pretty + *i* + *er* The yellow rose is prettier than the white one.
Pretty + *i* + *est* But the red one is the prettiest rose of all!

PRACTICE

Read each sentence. Circle the adjective that describes a noun. Then explain what the adjective tells about the noun.

Example: The (striped) shirt matches the jacket. ***Striped* tells what kind of shirt.**

1. Place the glass bowl on the table. ____________________
2. Hang the large signs by the doors. ____________________
3. Are the white tables arranged in a row? ____________________
4. We expect many salespeople to attend the meeting. ____________________
5. Cups and cold water will be placed on the tables. ____________________
6. The pots have fresh coffee. ____________________

Read each sentence and the adjective that follows it. Write the correct *-er* or *-est* form of the adjective to complete the sentence.

7. My aunt is one of the ____________ people you will ever meet. *kind*
8. Let's hang up the ____________ picture of the family instead of the later one. *early*
9. Today is the ____________ day we have had this year. *warm*
10. This is the ____________ loaf of bread I've ever baked. *tasty*
11. Ask the ____________ of the two children to help carry the box. *strong*
12. The stars shine most brightly in the ____________ skies. *dark*

Complete the sentence by adding an adjective that modifies the noun. Be sure the adjective you add makes sense.

13. Only the ____________ swimmer will win the contest.
14. These dishes are ____________ than those ones.
15. This band plays ____________ music than that band does.

LESSON 19: Pronouns and Antecedents

A **pronoun** is a word that takes the place of one or more nouns. The noun that the pronoun replaces is called the **antecedent**. The antecedent comes *before* the pronoun.

Examples:

The phone is on the counter.
It is on the counter.

Phone is a noun. *It* is the pronoun that replaces *phone*. So *phone* is the antecedent of the pronoun *it*.

Pronouns agree in *number* with the nouns they replace. A singular pronoun replaces a singular noun. A plural pronoun replaces a plural noun.

Examples:

The man waited on the train platform.
He waited on the train platform.

The noun *man* is singular, so the singular pronoun *he* replaces it.

Examples:

The men waited on the train platform.
They waited on the train platform.

The noun *men* is plural, so the plural pronoun *they* replaces it.

Pronouns also agree in *form* with the nouns they replace. The subject form of a pronoun replaces a noun used as a subject. The subject does the action in the sentence. The object form of a pronoun replaces a noun used as an object. It receives the action in the sentence. The object form often comes after the verb in a sentence.

Review the forms of subject and object pronouns.

Subject Pronoun	Object Pronoun
I	me
you	you
he	him
she	her
it	it
we	us
they	them
who	whom

Remember: The subject tells the person(s) or thing(s) the sentence is about. The object is the person(s) or thing(s) that receives action. Many times, the object pronoun comes *after* the verb.

Examples:

Jeans and T-shirts can be washed together. They can be washed together.

Object pronouns often follow **prepositions** like *at, about, for, from, of, to*, and *with*. Prepositions give more information about the relationship of the nouns and objects in the sentence.

Examples:

The children went *with* their families. The children went *with* them.

Who (subject pronoun) and whom (object pronoun) are pronouns used to ask questions. They match in number and form, too.

Examples:

Who tossed the jeans into the washing machine? He did.
To whom do these jeans belong? They belong to him.

PRACTICE

Read each sentence. See the underlined pronoun. Circle the pronoun's antecedent.

1. I will meet Jan and drive her to school.
2. The work crew fills the holes in the road and makes it smooth again.
3. Marcus, who is playing the guitar, is my brother.
4. Hal and Fran live next door, and they have been our good friends for years.
5. Ask the students about the assignment, and write down the information for them.

Read each sentence. Circle the pronoun that agrees with the underlined nouns in number and form.

6. Rosa can give us directions to the concert hall because (she, her) has been there before.
7. We bought grapes and oranges and packed (it, them) in the cooler for later.
8. The game went into overtime, but the fans stayed to watch (it, them) until the end.
9. When traffic is this heavy, people must be patient as (they, he) drive.
10. The dog and I had a good run in the park, but (we, us) got wet in the rain showers.

Write a subject or object pronoun to complete each sentence. The pronoun must agree in number and form with the noun it replaces.

11. We heard from Jason that the movie starts at 7:00 p.m.
    We heard from ________ that the movie starts at 7:00 p.m.
12. Jason and Tess will arrive early to buy tickets.
    ________ will arrive early to buy tickets.
13. They will buy tickets for their friends, too.
    They will buy tickets for ________, too.
14. How much does popcorn cost?
    How much does ________ cost?

LESSON 20: Declarative and Interrogative Sentences

A **declarative sentence** is a sentence that tells something. Declarative sentences are also called statements.

Use a period (.) at the end of a declarative sentence.

Examples:

The smell of mint or rosemary helps some people stay alert**.**
Lavender, on the other hand, helps some people relax**.**

An **interrogative sentence** is a sentence that asks something. Interrogative sentences are also called questions.

Use a question mark (?) at the end of an interrogative sentence.

Examples:

Do you like the smell and taste of peppermint**?**
Have you tried to grow herbs such as mint in pots**?**

Remember: Begin every declarative sentence or interrogative sentence with a capital letter.

Examples:

We grow mint and lavender in pots in the kitchen.
Would you like me to cut some fresh mint for you?

PRACTICE

Read each sentence. Decide whether it is a declarative sentence or an interrogative sentence. Underline the sentence type. Then add the correct end mark—a period or a question mark.

1. Do you pick up your mail every day______
   (declarative, interrogative)
2. I have to drive to the post office to get my mail______
   (declarative, interrogative)
3. Netta checks her locked mailbox once a week______
   (declarative, interrogative)
4. Is your mailbox locked, or can anyone open it______
   (declarative, interrogative)
5. I wonder why we don't get mail on Sundays______
   (declarative, interrogative)

Rewrite each sentence so that each begins with a capital letter and ends with the correct end mark—a period or a question mark.

6. how can you help your children do well in school
   ____
7. you can set up a quiet place for study and homework
   ____
8. it helps to have one folder where you keep all assignments and teacher information
   ____
9. do you attend parent-teacher conferences
   ____
10. watching your children learn and grow is an exciting experience
    ____

Read each interrogative sentence. Then write a declarative sentence that answers the question. Use the correct end mark.

11. Do you stretch and exercise daily?
    ____
12. Have you been sleeping well lately?
    ____

LESSON 21: Exclamatory and Imperative Sentences

An **exclamatory sentence** is a sentence that shows strong feeling. Exclamatory sentences are also called exclamations.

Use an exclamation point (!) at the end of an exclamatory sentence.

Examples:

That's exciting news!
I am so glad to meet you!

Some exclamations are so short that they are not complete sentences. These also end with an exclamation point. You can tell what these short exclamations mean from the context in which someone says them.

Examples:

Fantastic!
Oh, dear!

An **imperative sentence** is a sentence that gives an order or a direction. Imperative sentences are also called commands. The subject in imperative sentences is always "you." However, it is not said.

Use a period (.) at the end of most imperative sentences.

Examples:

Close the door.
Please have a seat.

Strong or urgent imperatives may end with an exclamation point.

Examples:

Stop!
Stand back!

Remember: Begin every exclamatory sentence and imperative sentence with a capital letter.

Examples:

Congratulations on your new job!
Turn left at the next green light.

PRACTICE

Read each sentence. Decide whether it is an exclamatory sentence or an imperative sentence. Underline the sentence type. Then add the correct end mark—an exclamation point or a period.

1. Close the door to keep the rain out_______
(exclamatory, imperative)

2. I can't believe you're finally here_______
(exclamatory, imperative)

3. Please listen carefully_______
(exclamatory, imperative)

4. Be careful on the icy sidewalk_______
(exclamatory, imperative)

5. Let's celebrate the team's win_______
(exclamatory, imperative)

Rewrite each sentence so that each begins with a capital letter and ends with the correct end mark—an exclamation point or a period.

6. look at that sunset

____

7. it's so beautiful

____

8. tell me why sunsets are colorful

____

9. please speak slowly

____

10. thank you for explaining so clearly

____

Write an exclamatory or imperative sentence that you might say in response to the situations. Use the correct end mark.

11. A friend shares exciting news with you.

____

12. People in the movie theater are talking noisily behind you.

____

Circle the correct answer.

1. Which group of words is a correctly written sentence?
   A. Marla drives to her children's school every morning
   B. marla drives to her children's school every morning.
   C. Marla drives to her children's school every morning.
   D. marla drives to her children's school every morning

2. Which group of words is a complete thought?
   A. The children outside.
   B. Waiting outside the school.
   C. The children outside the school.
   D. The children wait outside the school.

3. Read the sentence. Then find the subject.
   The crossing guard guides the children to the car.
   A. guides
   B. to the car
   C. The crossing guard
   D. the children

4. Read the sentence. Then find the predicate.
   Marla gives each child a hug.
   A. Marla
   B. Marla gives
   C. gives each child a hug
   D. a hug

5. Read the sentence. Then find the predicate.
   The family drives straight to soccer practice.
   A. family
   B. The family
   C. straight to soccer practice
   D. drives straight to soccer practice

Circle the conjunction that best connects the ideas.

6. School starts in two weeks, _______ we will buy new folders, pencils, and paper.
   A. and
   B. for
   C. or
   D. so

7. I am cleaning the flower beds and pulling weeds, _______ I can plant herbs and flowers.
   A. so
   B. for
   C. yet
   D. but

Circle the answer that best joins the sentences into a single clear sentence.

8. Today's weather is hot and humid.
   We promised to clean up the park.
   We will drink lots of water while we work.
   A. Today's weather is hot and humid, but we promised to clean up the park, so we will drink lots of water while we work.
   B. So today's weather is hot and humid, we promised to clean up the park, but we will drink lots of water while we work.
   C. We will drink lots of water while we work, yet we promised to clean up the park, so today's weather is hot and humid.
   D. We promised to clean up the park, so today's weather is hot and humid, nor will we drink lots of water while we work.

9. We do not want to get scratched. We wear work gloves. The bushes by the playground are thorny.
   A. We do not want to get scratched, nor do we wear work gloves, for the bushes by the playground are thorny.
   B. We wear work gloves, but we do not want to get scratched, yet the bushes by the playground are thorny.
   C. The bushes by the playground are thorny, for we wear work gloves, so we do not want to get scratched.
   D. The bushes by the playground are thorny, and we do not want to get scratched, so we wear work gloves.

Circle the correct answer.

10. Read the sentence. Which word in the sentence is an adjective that modifies a noun?

    Please help me find the brown suitcase.

    A. Please

    B. help

    C. brown

    D. suitcase

11. Read the sentence. Then circle the correct form of the adjective to complete the sentence.

    We serve the _______ iced tea in town.

    A. fresh

    B. fresher

    C. freshest

    D. freshier

12. Read the sentence. Then circle the correct form of the adjective to complete the sentence.

    The _______ woman of the two has interesting stories to tell.

    A. old

    B. older

    C. olderest

    D. oldest

Decide which pronoun replaces the underlined noun. Circle the correct answer.

13. Please ask <u>Marge</u> if _______ would like a glass of water.

    A. she

    B. her

    C. them

    D. it

14. My <u>mom and dad</u> are here, and _______ will drive the team to practice.

    A. she

    B. he

    C. them

    D. they

Circle the correct answer.

15. Choose the declarative sentence that is written correctly.

    A. repainting a room is a quick and easy way to freshen up your home

    B. repainting a room is a quick and easy way to freshen up your home.

    C. Repainting a room is a quick and easy way to freshen up your home.

    D. Repainting a room is a quick and easy way to freshen up your home?

16. Choose the interrogative sentence that is written correctly.

    A. How many pears have you picked from this tree?

    B. How many pears have you picked from this tree.

    C. How many pears have you picked from this tree!

    D. how many pears have you picked from this tree?

17. Circle the exclamatory sentence that is written correctly.

    A. What a fantastic day!

    B. What a fantastic day.

    C. what a fantastic day!

    D. What a fantastic day?

18. Read the sentence.Then circle the answer that tells what kind of sentence it is.

    Grab your umbrella and raincoat.

    A. declarative

    B. exclamatory

    C. imperative

    D. interrogative

LESSON 22: Predict

You can use the skill of **predicting** to help you better understand what you read. When you predict, you use what you know and what you read. You put them together to make a smart guess about what an author will say next.

You read more actively when you read, pause, predict, and read on. Follow these steps to predict as you read.

- Read part of a text. You might read just the title, or the title and first sentences.
- Pause and think. What do you know about the subject? How can you connect what you know to what you just read?
- Predict. Based on your answers, what do you think the author will say next?
- Check your prediction by reading the text. Stop. Did you guess correctly?

Then read on, and repeat the steps. You can take notes as you read and make predictions.

Some predictions will be correct. Some won't. That's okay. Just update your predictions with the new information you have read.

PRACTICE

Use this passage to answer each question. Do not read the passage before you begin. Follow the directions in order. Do not skip ahead to other questions.

Costs of the Common Cold in the United States

1 The common cold is a big problem. People catch about 500 million colds a year! Drugs and doctor visits cost a lot of money. Companies lose money when people are sick and miss work. Less work gets done.

Lost Work Time

2 Colds cost the United States $40 billion a year. Half of this cost is due to missed days of work. People stay home when they are sick. Children miss school when they are sick. Parents stay home to care for sick children.

Medical Costs

3 A cold comes with a large price tag. People spend close to $8 billion on doctor visits. They spend $3 billion on over-the-counter drugs. About $400 million is spent on prescription drugs. About $1 billion is spent on antibiotics. Viruses cause most colds. Antibiotics do not cure viruses.

4 Only time cures the common cold. If we could find a better cure, we could save billions of dollars a year.

1. Read the title. Think about what you pay for when you have a cold. Predict what "costs" the author might talk about. Write your prediction.

____

2. Now read paragraph 1 of the passage. What kinds of "costs" does the author talk about?

____

Maybe you need to change your prediction. That's okay.

3. Read the first subtitle. A subtitle will tell you the topic or what a section is about. How does the subtitle help you predict what you will read about in this part of the passage?

____

4. Read the second subtitle. Predict the kind of information you will find in this section.

____

5. Read paragraph 4, and then read the prediction you wrote in question 4. Did you guess correctly? How did the subtitle help you predict the information in this part of the passage?

____

6. Think about your experience with having a cold. What information do you already know about the medical costs of colds?

____

7. Reread the amounts of money spent on colds. Predict how finding a cure for the cold would change these amounts.

____

8. The last sentence says, "If we could find a better cure, we could save billions of dollars a year." Think about what you have read in the passage so far. Make a prediction about what the author might talk about next.

____

LESSON 23: Scan

Often, you read a text slowly and carefully. You want to read and understand every sentence. But sometimes you need to read quickly. You want to know only a few pieces of information. Reading over a text quickly to find specific information is called **scanning**. You might **scan** a text for these purposes.

- to review information you studied in the past
- to find important details about a subject
- to get just enough information to act or to make a decision

Some types of text are easy to scan.

- Schedules are laid out by hours and days. Find the time or day you need to know about. Scan just that part of the schedule.
- Lists are often arranged alphabetically or in groups. Find the letter of a word or the group name you need to know about. Scan just that part of the list.
- Ads often have words and images. If you need to know costs or dates, scan for numbers. If you need to know about a product, scan for details about the product and look at the pictures.

Paragraphs have many lines of text. They are harder to scan. Before you scan a text with many paragraphs, decide what information you want to find. Think of keywords—words that name the information you need.

Example:

You see a passage called "Training Your New Puppy."
You need information about teaching a puppy to walk on a leash.
Scan for keywords such as *leash, walk, heel,* and *sit.*
Read the paragraphs that include these words. Skip the rest of the article (for now).

Hint: If you read a passage online, use the Find function to scan for keywords. A Find box will appear at the top or bottom of your screen. Move your cursor and click inside the box. Type the keyword you want to find. Click *Enter* or *Return*. This will search the web page for the word you typed.

PRACTICE

On the next page is a passage about octopuses. Answer questions 1–3 before you scan the passage.

1. Imagine you want to know how big octopuses can grow. Should you scan for numbers or capitalized words?

____

2. Imagine you want to know how octopuses catch food. Circle the words you would scan for.

prey   color   food   tiny   bite

3. You want to know what dangers octopuses in the ocean face. Write three words that you could scan for to find this information.

____

Octopus Facts

1 The octopus lives in the ocean. The biggest and smallest ones live off the west coast of the United States. One type of octopus can weigh 100 pounds. It can be 30 feet long. Another is smaller than an inch.

2 An octopus protects itself from enemies. If an enemy sees it, the octopus lets out a cloud of black ink. This cloud helps hide the octopus. It can then swim away. The octopus changes its skin color, too. Each skin cell has three colors that it can mix. The octopus changes its color until it blends in with its surroundings. It can let out ink many times before its ink sac is empty. The ink can kill the octopus. It must quickly leave the area, or it might die.

3 Many octopuses bite their prey. This poisons them. Octopuses use their sharp beaks to bite through the shells of crabs and lobsters. An octopus with a small beak can bite through a diver's wetsuit!

Answer each of the following questions about the passage.

4. Which paragraph contains facts about octopus size? __________
5. Which paragraph uses the word for an animal that another animal catches to eat? __________
6. In your own words, tell what you have learned about how octopuses face danger.

____

Answer the questions by scanning the car ads below.

'12 Ford Taurus. Loaded, every added feature. Excellent condition. $10,500. Call 788-9208.

'02 Ford Thunderbird. Maroon, 146,000 miles. Good condition. Runs but needs engine work. $1,750 or best offer. Call 789-0081.

'11 Honda Civic. 2-door, keyless entry, power everything. Silver with gray interior. Very sporty! $10,800. Call 768-1274.

'08 Ford Focus. Perfect condition! Air conditioning. 5,500 miles. $9,500. Call 890-2641.

'10 Ford Fiesta. Excellent condition. 74,000 miles. Pay off loan of $5,000 or take over payments of $225/month. Call 930-7823.

Imagine that you are looking for a used car. Answer the questions about the car ads.

7. You like Fords. Scan for the word *Ford*. How many cars can you read about? __________
8. You live in a city where summers are very hot. What words should you scan for if you want to stay cool?

____

9. You want to know how much a car has been driven. Write a word you can scan for to find this information.

____

10. Cost is most important to you. What symbol can you scan for to find out a car's cost? __________
How much does the Ford Focus cost? __________
11. Scan for color words. Which cars' colors are listed? __________
12. How can you locate the newest car? __________

LESSON 24: Skim

To **skim** a text is to read quickly to find its **main ideas**, or main point of the reading. When you skim, you can underline main ideas or take notes that state the ideas in your own words. When you finish skimming a text, you have an outline—a list of the main ideas in the order you see them in the text.

How are skimming and scanning different?

You *skim* to get the "big picture" of the text. Skimming helps you find each main idea and see how the main ideas fit together. You skip the details in the text.

You *scan* to find just one or two kinds of information in a text. You read only the parts of the text that have that information. You skip the rest of the text.

Follow these steps to skim a text.

- Read the title. The title often tells the subject. For example, a passage called "The Right Time to Buy Electronics" will likely tell when and where to buy products like computers and televisions.
- Read the first paragraph. This is the introduction. Often, the last sentence in this paragraph contains the main idea. This is what the rest of the reading will be about. Underline the main idea or write it in your own words.
- Read the first sentence of each of the following paragraphs. The first sentence is often the **topic sentence**. It usually states the main idea of the paragraph. Underline this idea or write it in your own words.
- Read the last paragraph. This is the conclusion. The text's main ideas are often restated in the conclusion. Underline the main ideas or write them in your own words.
- Pay attention to subtitles. Like titles, subtitles often tell you the subject of the part of the text that follows. They also provide clues about how the ideas will be organized.
- Skip sentences that start with "For example" or "For instance." These sentences contain supporting details, not main ideas. Skip the details when you skim.

Remember: When you skim, you read parts of the text to get the main ideas. Later, you may want to read the whole text and get the details, too.

PRACTICE

Skim the following passage. Then answer the questions.

Children Learn from Allowances

1 Should children get a regular allowance? If so, how much is okay? An allowance gives children more than spending money. It teaches them valuable lessons. They learn to save their money. They learn to spend it wisely.

Don't Pay for Chores

2 Should children be paid an allowance for doing chores? Most experts say no. They think children should do chores because they are part of the family, not because they are getting paid. As members of the family, they should help the family meet its needs.

How Much? How Often?

3 Parents can match the amount of an allowance to the child's age. A child between the ages of six and nine might receive $2 to $6 a week. A child ages nine to twelve might get $6 to $10 a week. Consider how much your family makes. Think about what a fair amount is to give your child.

4 A child may spend all her allowance before the end of the week. This is a good chance to learn a lesson. As Ben Franklin, an early American writer, once said, "When the well's dry, we know the worth of water." Help your child learn from her actions. Do not give her the next allowance early. This will help her appreciate her allowance. Maybe she will learn to spend it more wisely.

5 Ben Franklin had more advice. He said, "An example is the best sermon." If we manage our money wisely, our children will follow our example. If we often spend all of our money before the next payday, our children will do the same.

1. Paragraph 1 begins with a question. Review the title and the paragraph. Underline the sentence(s) that answer the question. What is the answer to the question asked in paragraph 1, yes or no?

____

2. Skim paragraph 1 for two reasons parents should give children allowances. Write the reasons in your own words.

____

3. In your own words, write the main idea you found as you skimmed these paragraphs.

   Paragraph 2 main idea:

   ____

   Paragraph 3 main idea:

   ____

   Paragraph 4 main idea:

   ____

4. How does the subtitle "Don't Pay for Chores" help you prepare to skim paragraph 2? Think about keywords.

____

5. How does the subtitle "How Much? How Often?" prepare you to skim paragraphs 3, 4, and 5?

____

6. Which sentence is the main idea in paragraph 5? Write the sentence.

____

7. Review the main ideas you underlined or wrote as you scanned the passage. Then write one sentence that sums up the main idea you learned from scanning the passage.

____

LESSON 25: Infer

To **infer** is to connect something you know and something you read. Once you have made the connection, you can come up with a new idea based on the connection.

This process is also called inferring, or making an inference. Sometimes an inference is called an educated guess.

You make inferences every day. Any time you connect what you know with new information and figure things out, you are inferring.

What you already know + New information you read = Inference

Imagine that you meet a good friend. The friend's face looks sad. His shoulders are down, and he's not standing up straight. You know your friend is usually cheerful and full of energy. You connect what you know to what you see. You infer that your friend is sad today for some reason. Now you can act based on what you've just figured out.

When you read, you use the same process. You infer what a new word means by connecting it to what you already know. You infer what motivates people in nonfiction texts and characters in stories by connecting what they do and say to your real-world experiences. You infer what an author thinks or feels but does not say outright. You infer to connect how what you read connects to the world around you.

Some people talk about "reading between the lines" of an article or story. They mean that readers think about what the author states and what the author suggests. That's inference.

Sometimes your inferences will turn out to be right, but not always. It's okay to make an incorrect inference now and then. You can correct your guesses as you get more information.

PRACTICE

Read the opening paragraph of a story. Then answer the questions.

> Kay stared at the wallpaper in her grandmother's bedroom. Grandmother's funeral had been last week. Now Mom was making Kay help her clean the old house. She kept calling Kay to come into the kitchen. Mom wanted Kay to see another "treasure" she had found. Kay frowned. All she could think about were the times when her grandmother would criticize and scold her.

1. Notice how Kay acts. Based on her thoughts and actions, what can you infer about how she feels about her grandmother?

____

____

2. Read the these sentences again.

> She kept calling Kay to come into the kitchen. Mom wanted Kay to see another "treasure" she had found.

Does Kay think the things her mother shows her are treasures? What makes you think so?

____

Read the next part of the story. Then answer the questions.

> Kay pulled open the top drawer of a dresser. Scarves and paperback books flew out of it. She couldn't believe how unorganized her grandmother had been. Kay smiled. Many years from now, she imagined that her own granddaughter might clean out Kay's room. Would this girl be surprised when all the stuff Kay saved flew out of her dresser?
>
> Then Kay found her own school picture from last year. On the back of it, her grandmother had written: "Here is my darling Kay, the love of my life. She looks just like Rachel at this age." Kay gently set the picture to one side. This was a keepsake she truly would treasure for years to come.

3. What does Kay do that suggests a change in the way she feels about the task of cleaning out the old house?

____

4. Kay makes a connection between her grandmother's habits and her own habits. Use what you know about people's relationships to infer how this connection makes Kay feel.

____

____

5. Remember what you inferred about Kay's relationship to her grandmother from the first paragraph. What were Kay's memories mostly about?

____

____

6. Why might Kay be surprised by what is written on the back of the picture? Use details from the story to support your inference.

____

____

7. What can you infer by the way Kay sets the picture to one side?

____

____

8. Describe how Kay's character changes from the beginning of the story to its end. Your description is an inference based on many details and on what you know about people.

____

____

____

Scan the schedule and answer the questions. Circle the correct answer.

Porter Airline Service: Baltimore, MD, to Portland, OR

Flight Number	Departs from Baltimore, MD	Arrives in Houston, TX	Departs from Houston, TX	Arrives in Portland, OR	Total Travel Time
1721	7:10 a.m.	10:01 a.m.	10:35 a.m.	2:11 p.m.	10 hours, 1 minute
509	7:34 a.m.	—	—	11:47 a.m.	7 hours, 13 minutes
465	8:38 a.m.	10:28 a.m.	11:07 a.m.	3:33 p.m.	10 hours, 55 minutes
2651	9:30 a.m.	11:34 a.m.	1:30 p.m.	5:40 p.m.	11 hours, 10 minutes
7582	3:35 a.m.	—	—	7:56 p.m.	7 hours, 21 minutes

1. Which columns do you scan to find the shortest flight from Baltimore, MD, to Portland, OR?
   - **A.** Flight Number and Departs from Baltimore, MD
   - **B.** Arrives in Houston, TX, and Departs from Houston, TX
   - **C.** Arrives in Portland, OR, and Total Travel Time
   - **D.** Flight Number and Total Travel Time

2. Which is the shortest flight from Baltimore, MD, to Portland, OR?
   - **A.** Flight 509
   - **B.** Flight 465
   - **C.** Flight 2651
   - **D.** Flight 7482

Circle the correct answer.

3. What reading strategy asks readers to connect what they know and what they read to guess what comes next in a text?
   - **A.** infer
   - **B.** predict
   - **C.** scan
   - **D.** skim

4. An article has the title "Make the Most of Your Summer Vacation." Which prediction about the article is most likely correct?
   - **A.** The article tells where to go on vacation.
   - **B.** The article tells how to get to a vacation spot.
   - **C.** The article tells how to use vacation time well.
   - **D.** The article tells how to pay for a vacation.

5. What reading strategy asks readers to look through a text quickly and find information about a specific subject?
   - **A.** infer
   - **B.** predict
   - **C.** scan
   - **D.** skim

6. You find a long report called "The Best Schools in the City." You need information about schools for older students. Which keyword(s) should you scan for?
   - **A.** preschool
   - **B.** high school
   - **C.** charter school
   - **D.** elementary school

7. What reading strategy asks readers to read a text quickly and find the main ideas?
   - **A.** infer
   - **B.** predict
   - **C.** scan
   - **D.** skim

8. When you scan a text, what can you skip, or not read?
   - **A.** main ideas
   - **B.** title and subtitles
   - **C.** the first paragraph
   - **D.** examples and details

9. What type of text is not easy to scan?
   - **A.** chapters
   - **B.** ads
   - **C.** schedules
   - **D.** lists

10. You read a story. You think about what you know, and you make a connection to come up with a new idea.
    What reading strategy are you using?
    - **A.** predict
    - **B.** infer
    - **C.** scan
    - **D.** skim

Read the following paragraph. Circle the correct answer.

> Kay cleaned out all the drawers in the chest except the bottom drawer. She opened it. She saw that it was full of cards and old photos. Kay almost cried when she found two cards tied together with a red ribbon. They were valentines. Her grandparents had given each other these cards. The cards were signed with love. She also found pictures of her grandparents together. They looked very young. There also were many pictures of her mom Rachel and Uncle Scott.

11. Which reading strategy is used to write this sentence?

    > Kay's grandmother kept many cards and photos that were special to her.

    - **A.** infer
    - **B.** predict
    - **C.** scan
    - **D.** skim

12. Which reading strategy is used to write this sentence?

    > I think that Kay will choose and keep some cards and photos for herself.

    - **A.** infer
    - **B.** predict
    - **C.** scan
    - **D.** skim

LESSON 26: Understanding Formatting Clues

Texts in print and online often have **formatting clues**. Formatting clues draw your attention to important ideas and words. Formatting clues also help you gather information.

A title tells what the text is about. Headings help divide the text and help readers find specific information. Bulleted lists show the most important parts. Italics or bold print draw attention to important words or phrases.

Study the text about a group that rates games. Notice the formatting clues that are used.

Example:

title → **Video Game Rating System Helps Parents Choose Games**

heading → **What is the ERSB?**
ESRB stands for the Entertainment Software Rating Board. It is a group that rates video games and apps. The group's task is to look at game content. It also suggests how old players should be to play a game.

heading → **What is in each rating?**
Each game or app has a rating. The rating has three parts.

italicized items →

Rating system: This is a letter assigned to games. It tells for what ages the game is okay.

Content descriptors: This part tells why a game got its rating. It might point out things in the game that would interest or worry someone.

Interactive elements: This part tells how a game or app connects to other games and to the Internet. It tells if the game stores or shares a player's information. It tells if the game lets players buy things while they play the game. It tells if the game lets players talk to other players while they play the game.

heading → **What are the rating categories?**
Rating categories are quick ways for parents to choose games. Ratings give an age range for players of the game.

bulleted list →

- EC — Early childhood—the game is okay for young children.
- E — Everyone—the game is okay for all ages. The game may have mild cartoon violence and mild language.
- E 10+ — Everyone ten years of age and up—the game may have cartoon violence and mild language.
- T — Teen (13 years and up)—the game may have some violence, crude humor, some strong language, and some suggestive language.
- M — Mature (17 years and up)—the game may have violence, blood and gore, and strong language.
- A — Adults only (18 years and up)—the game may have a lot of strong violence and gore, adult content, and a lot of strong language.

heading → **You are the parent!**
The ERSB gives information and guidelines. It is up to you to choose games for your child. First check the game's rating. Then read the content descriptions. Go online to find more about the game. Then choose the games your child is ready for.

PRACTICE

Answer the questions about the formatted text on the previous page.

1. A reader does not know what ERSB stands for. Under which heading will the reader find information about the ERSB? ______
2. The text has three terms in italics. How are the three terms connected to one another?

   ______
3. Skim the bulleted list. In what order are the items on the list arranged?

   ______
4. Reread the "You are the parent!" section. What information is found in this section?

   ______

Read the rating information for a game. Then answer the questions.

Parent Review: *Let's Play with Pets!*

Description

Let's Play with Pets! is a new video game. It is made for young children. The game has children set up a pet store. Children can then "adopt" pets—cats, dogs, birds, and bunnies—to care for. Children must feed pets on a schedule. They can play with pets by throwing them balls or taking them for walks. The graphics are colorful and bright. The pets are cute and fluffy. They have large eyes and happy faces.

Rating—Pros

The ERSB has rated this game *EC—early childhood.* We agree with this rating. The content is okay for young children. It is easy for children to understand how to play the game. And the game teaches them that pets must be cared for.

Interactive Elements—Cons

The ERSB points out that this game allows in-game purchases. The game lets children "share" their pets with other players. The ERSB is concerned about some interactive parts of the game.

- Children may not understand that when they "adopt" some pets, they must pay for them. The game does not tell children to check with parents before buying a pet.
- Children may not understand that they may be "sharing" pets with adults pretending to be children while playing the game.

These interactive elements do too little to protect children from inappropriate use of the game.

5. Which section of the text presents problems with the game?

   ______
6. What information is in the bulleted list?

   ______
7. What do the reviewers like about the game? In what section is this information?

   ______

LESSON 27: Main Idea

Each paragraph you read has a **main idea**. The main idea is the most important idea in the paragraph. It is what the author would like you to remember later. The main idea is supported by details. Details give more information about the main idea.

Read the paragraph about monarch butterflies. As you read, ask yourself, "What is this paragraph mostly about?"

Example:

Monarch butterflies fly a lot during their lives. When it starts to get cold, millions of them fly south. They go from the United States and Canada to Mexico. They spend the winter in trees in the mountains of Mexico. In the spring, they fly back north.

The paragraph mostly tells about monarchs' journey. The main idea is in the first sentence, *Monarch butterflies fly a lot during their lives.*

The other sentences in the paragraph give details, or other specific information, about monarchs' journeys. The details tell where they fly, where they stay in winter, and when they fly back. None of these details on its own tells what the paragraph is mostly about. They all support the main idea by giving details.

Read another paragraph about monarchs. As you read, ask yourself, "What is this paragraph mostly about?"

Example:

Scientists keep track of the monarchs. They want to know how many of them make the trip each year. In Mexico, the monarchs live in tree branches. There are millions of them in the trees. To count them, scientists measure the size of the forest where they live. They usually take up about twenty-four acres of forest each winter.

Find the main idea in the paragraph. The main idea is in the first sentence, *Scientists keep track of the monarchs.*

The other sentences have details that tell how scientists keep track of the butterflies. None of these sentences, on its own, tells the main idea.

Did you notice that the main idea in each paragraph is in the first sentence? You will often find a paragraph's main idea in the first sentence. But not always! Sometimes the main idea comes later or even in the last sentence.

PRACTICE

Read more about monarch butterflies. Then answer the questions.

1 Why do monarchs fly south in the winter? They go to the mountains in Mexico to find better weather. They need to get out of the cold. The area there is not too hot or too cold. It is not too wet or too dry.

2 Sometimes the monarchs have a bad year. In January 2002, a storm hit the mountains. It killed more than 200 million monarchs. Before the winter storm, the monarchs lived on about twenty-four acres. After the storm, they lived on only about five acres. The ones that made it through the storm were lucky. They could fly north in the spring. They could lay eggs. They could continue their life cycle. But would there be enough of them?

3 Scientists worried that the storm may have killed too many monarchs. The trip is hard. Many butterflies die along the way. If enough of them did not make it, then that could be the end of the monarchs.

4 The next winter, though, millions of them came back to the mountains in Mexico. They took up twenty acres of forest. That was only four acres fewer than before the storm. The storm had not caused the end of monarchs.

5 Scientists were surprised. They did not think so many monarchs would return. The numbers were amazing!

1. What is the main idea of paragraph 1?

____

2. Paragraph 2 tells about an event and its results. What was the event?

____

3. What is the most important idea about the event described in paragraph 2? Write this main idea in your own words.

____

4. Reread the sentences below from paragraph 3. Write *Main Idea* or *Detail* to describe each sentence.

a. Scientists worried that the storm may have killed too many monarchs. ______________

b. The trip is hard. ______________

c. Many butterflies die along the way. ______________

5. The last paragraph is very short—just three sentences. In your own words, write what this paragraph is mostly about.

____

LESSON 28: Supporting Details

Supporting details help you to better understand the main idea. The main idea tells what the paragraph is about. Supporting details provide information to prove, explain, or make the main idea clearer. They can help answer readers' questions about the main idea.

Supporting details can give several useful kinds of information. **Signal words** let the reader know how the details support the main idea.

- Details can explain why the main idea matters to readers. Look for signal words like *because, so that, as a result,* and *then*.
- Details can explain how the main idea works. Look for signal words that show order like *first, next, then,* and *finally*.
- Details can give examples of the main idea. Look for signal words like *for example, for instance, also,* and *in fact.*
- Details can explain why something happens. Look for signal words like *as a result*, *if so*, *due to*, and *for this reason*.
- Details can give facts and numbers about the main idea. Look for measurements, numbers, amounts, and percentages.

When you read a paragraph, find the main idea. Then find the supporting details. Ask yourself, "How do these details develop the main idea? How do they help me understand the main idea better?"

Read the paragraph below. The main idea is underlined. The rest of the sentences are supporting details.

Example:

What kind of machines and tools do you think of when you hear the word *farming*? Do you think of a farmer with a tablet in a field? Or do you think of a farmer on a tractor? Modern farmers can be found doing both of those things. <u>Farmers today use computers just as much as they use other tools.</u>

In this paragraph, the supporting details come first. The main idea comes last. The author thinks that readers will be surprised to think of farmers as people using technology. The author puts the main idea last as part of that surprise.

PRACTICE

Answer the questions about the paragraph. The main idea is underlined.

On a farm now, the computer is as important as the plow. One person on the farm takes care of the business tasks. That person uses farm accounting software. This keeps track of money earned and spent. He or she checks market trends. He or she also looks at prices online to find the best price for tools or supplies. He or she emails or texts other farmers and companies.

1. Read the sentence below. How does this detail support the main idea of the paragraph?

   This keeps track of money earned and spent.

   ____

2. Name three ways farmers use computers to manage a farm, besides using accounting software.

   ____

   ____

   ____

Answer the questions about the paragraph. The main idea is underlined.

On a modern farm, technology helps farmers make more money. In a field of corn, it is important that every plant sprouts at the same time. For instance, a seed may be planted too deeply. If so, the stalk appears later than others. The tall stalks nearby will shade the small plant from light. The small plants will take more food from the soil. They will not grow a full ear. These late, small plants are called "nubbins." There may be more close to 1,000 of these "nubbins" per acre. This is about six bushels per acre. Imagine you are a farmer with 500 acres of corn, and corn sells for $4 per bushel. Those six bushels of nubbins make the farmer lose $12,000. Farmers can use machines to plant seeds. These machines make sure that seeds are planted at the same depth. This helps plants sprout at the same time and avoid nubbins.

3. Skim the supporting details. Write the sentence with signal words that give an example.

   ____

4. Read the list of types of details on the previous page. What kinds of details are in this paragraph?

   ____

5. How do the details in this paragraph support the main idea? How do the details help you understand the main idea better?

   ____

   ____

LESSON 29: Fact or Fiction Elements

When you read texts that inform and report, you look for **facts**. Facts are information that you can prove. You can go to sources to check whether the facts are correct. For example, texts that inform often include facts that answer four basic questions.

Basic Questions	**Facts**
Who was there?	Names of people, number of people
What happened?	Description of activities
When did it happen?	Times and dates
Where did it happen?	Names of buildings, places, towns, cities, countries

Sometimes you will read texts that have factual elements and fictional elements. Then you must decide which parts of the text are fact and which parts are fiction. Here are some types of fictional elements.

- Figurative descriptions. These are details that use images to paint a picture, such as "the sunset sky blazed with fiery oranges and reds."
- Exaggerations for effect. An example is "a million little fireflies seemed to blink in the night."
- Exaggerations for humor. An example is "we could almost go swimming in the humid air."

These fictional elements add interest and humor to writing. Readers should not confuse them with facts.

PRACTICE

Read the post from a blog. Then identify factual elements and fictional elements from the post.

HOME ABOUT BLOGS CONTACT

The Wet Blankets CD Release Party, February 2, 2016, at Arlo & Annie's

Local band The Wet Blankets showed they know how to host an event at last night's release party. They sang songs from their new CD, *Flop-Eared Hound Dog*. The folk duo packed the house! Rivers of fans flowed in to listen and dance. They bought CDs and T-shirts. Lenny and Arletta took breaks between sets to visit with fans and sign CDs.

"Arletta sang 'Heartbroken Honey'," one fan reported. "There wasn't a dry eye in the room. She really knows how to sing those sad, sad songs."

Flop-Eared Hound Dog is the third CD for The Wet Blankets. It is their first time to get a national tour. "The release party had us over the moon. What a way to start our tour!" Lenny said. "Arletta and I know that our fans are the best on the planet."

Fans can keep up with the tour on this blog. Lenny and Arletta plan to post a million pictures from their travels. This way all their fans can come along for the fun. First stop on the tour is Atlanta!

Decide if the text from the blog is a factual or fictional element. Write the text in the correct box.

Lenny and Arletta plan to post a million pictures from their travels.
There wasn't a dry eye in the room.
The release party had us over the moon.
Lenny and Arletta took breaks between sets to mingle with fans and sign CDs.
Flop-Eared Hound Dog is the third CD for The Wet Blankets.
Rivers of fans flowed in. . . .
Fans can keep up with the tour on this blog.
First stop on the tour is Atlanta!
. . .our fans are the best on the planet.

Factual Elements
Information that can be checked and confirmed

Fictional Elements
Language that describes things colorfully or that overstates for effect

Circle the correct answer.

1. How do headings help readers?
   - **A.** They signal the end of the text.
   - **B.** They tell what a section is about.
   - **C.** They tell what the whole text is about.
   - **D.** They draw attention to a word.

2. What is the purpose of this bulleted list?

> Video games for young children should include
> - Easy instructions
> - Bright colors
> - Limited interactive elements
> - Built-in pauses

   - **A.** It lists new games parents can buy for young children.
   - **B.** It explains how playing games can be helpful to young children.
   - **C.** It describes games that are appropriate for young children.
   - **D.** It sets limits on how long young children should play games.

3. Why is the word *any* formatted in italics in this bulleted list?

> Video games for young child should not include
> - *Any* violence, even funny or mild violence
> - *Any* strong language
> - *Any* scary themes

   - **A.** to show that the word is repeated
   - **B.** to stress the importance of the word
   - **C.** to make it easy and fast to read the list
   - **D.** to show how the items are different

4. What question does the main idea in a paragraph answer?
   - **A.** Why should I read this paragraph?
   - **B.** What is this paragraph mostly about?
   - **C.** What knowledge do I bring to this paragraph?
   - **D.** How does this paragraph fit with the rest of the text?

5. Which signal word(s) would be used for a supporting detail that gives an example?
   - **A.** next
   - **B.** so that
   - **C.** because
   - **D.** for instance

6. What do the signal words *As a result* tell you about a supporting detail?
   - **A.** It gives an example.
   - **B.** It shows a sequence of steps.
   - **C.** It tells why something is important.
   - **D.** It explains why something happens.

Read the paragraph. Then circle the correct answer.

> Milkweed plants are the best food for monarch caterpillars. Milkweed leaves can poison some animals. But caterpillars can eat the leaves. The poison does not hurt them. It makes them poisonous. A bird might try to eat a monarch caterpillar. But the bird will spit it out. It will not try again!

7. Which sentence states the main idea of the paragraph?
   - **A.** Milkweed plants are the best food for monarch caterpillars
   - **B.** The poison does not hurt them.
   - **C.** It makes them poisonous.
   - **D.** A bird might try to eat a monarch caterpillar.

8. How do the details in the paragraph support the main idea?
   - **A.** The details explain why other caterpillars avoid milkweed.
   - **B.** The details list birds and mammals that want to eat caterpillars.
   - **C.** The details give examples of other insects that eat monarch caterpillars.
   - **D.** The details explain why milkweed is a good food for monarch caterpillars.

Read the paragraph. Then circle the correct answer.

(1) Hydroponics is a way to grow plants with no dirt. (2) It uses less water than growing plants in dirt. (3) Water costs a lot. (4) Plants must have water to grow. (5) Most ways of watering plants waste water. (6) Not all the water gets to a plant's roots. (7) The rest of the water is lost. (8) Or the water soaks into the dirt. (9) With hydroponics, plants are often put in containers. (10) The containers have no dirt. (11)They may have sand or wood chips instead. (12) Water lines give just enough water to each plant. (13) Hydroponic plants also can be grown in greenhouses. (14) As a result, farmers can control every drop of water.

9. Which sentences explain why most ways of watering don't deliver all the water to a plant's roots?

A. Sentences 1–3

B. Sentences 5–8

C. Sentences 9–11

D. Sentences 11–14

Read the review of a new CD. Then circle the correct answer.

Country-Folk Duo Breaks Out with New CD

This month, I have a few words to say about *Flop-Eared Hound Dog*. It is the third CD from Lenny and Arletta of The Wet Blankets. This husband-and-wife team has many hometown fans. This CD, I think, will explode them onto the country music scene. The melodies sparkle. I can't stop singing the catchy songs. Most of the songs are new. But this country boy's heart just about burst with happiness over Lenny's version of that sweet old tune "The Tennessee Waltz." Be sure to catch Lenny and Arletta on tour. You owe it to yourself to go. I'm glad I caught their show when they came through Atlanta!

The Wet Blankets' new CD is available online and in stores.

10. Which detail from the review is a factual element?

A. Most of the songs are new.

B. This CD, I think, will explode them onto the country music scene.

C. But this country boy's heart just about burst with happiness. . . .

D. You owe it to yourself to go.

11. Which detail from the review is not a factual element?

A. This husband-and-wife team has many hometown fans.

B. This CD, I think, will explode them onto the country music scene.

C. It is the third CD from Lenny and Arletta of the Wet Blankets.

D. The Wet Blankets' new CD is available online and in stores.

12. Which detail from the review is a fictional element that uses figurative description?

A. This month, I have a few words to say about *Flop-Eared Hound Dog*. . . .

B. The melodies sparkle.

C. Be sure to catch Lenny and Arletta on tour.

D. The Wet Blankets' new CD is available online and in stores.

13. Which detail from the review is a fictional element that uses exaggerated language?

A. Be sure to catch Lenny and Arletta on tour.

B. The Wet Blankets' new CD is available online and in stores.

C. This husband-and-wife team has many hometown fans.

D. But this country boy's heart just about burst with happiness. . . .

LESSON 30: Signal Words in a Biography

A **biography** tells the story of a person's life. In a biography, a writer does not include every detail of a person's life. Instead, the writer chooses important events and details. These details give readers a certain view of the person.

A biography may answer these questions about the person.

- What was the person like?
- Why was the person important?
- What problems did the person face?
- Who helped the person?
- What can readers learn from the person about how to live well?

When you read a biography, you can ask these questions and look for their answers. The answers are often the main ideas of the writing.

Biographies are often arranged in time order. The writer uses **signal words** that help readers follow the events over the years of the person's life. Time-order signal words give clues about when something happened. Here are some time-order signal words.

Time-Order Signal Words		
after	finally	since
after a while	first	since then
at last	following that	soon
at the beginning	in the end	soon after
before	last	then
earlier	later on	to start with
early on	next	when

Biography writers may also use cause-and-effect signal words. These words can help writers explain why a person did certain things during his or her life.

Cause-and-Effect Signal Words		
accordingly	due to	so that
as a consequence	for this reason	therefore
as a result	if . . . then	this led to
because	in order to	thus
but	nevertheless	unless
consequently	results in	without

PRACTICE

Read the biography. Then answer the questions.

1 Amy Tan was born in 1952. She grew up in California. She had two brothers. Both of her parents came to the United States from China. Tan was born in the United States. She was the only Chinese American girl in her classes from third grade to high school. As a result, she often felt that she did not fit in. She tried to be more "American."

2 Tan's older brother and her father died of brain tumors when she was fifteen. The family wanted to get away. They moved to Europe. Tan came back to the United States to go to college.

3 In 1969, she went to college. Her mother wanted her to be a doctor. At first, she studied medicine. Then she changed her mind. She wanted to study English. She wanted to be a writer. Tan's mother did not speak to her for six months. Her mother was mad because of the change. Tan began to write stories after college. She put the stories together in a book.

4 Tan's mother became sick while she was writing her book. Tan promised to take her mother to China after she got well. She kept her promise. Tan's first book was published when she was in China. This book is called *The Joy Luck Club*.

5 Tan also wrote children's books. One is *The Moon Lady*. It is a story about a grandmother who tells a story. The grandmother tells her granddaughters about her scary adventures.

6 Amy Tan still writes books today. Most of her writing is about China and Chinese Americans. Her family and culture have played important roles in her life and in her books.

1. Read paragraph 1 again. What cause-and-effect words does the writer use?

2. Read paragraph 3 again. List the signal words that help readers follow the order of events in Amy Tan's life.

3. What problems did Amy Tan face?

4. Based on what you know about her life, what do you think Amy Tan is like ?

LESSON 31: Context Clues in Consumer Texts

Consumer texts are part of daily life. These texts tell about the parts of a product. They tell how to use a product safely. They tell how to act in a certain situation. They tell how consumers can get more information. Here are some examples.

- Manuals
- Safety booklets
- Food labels
- Signs on roads and businesses
- Internet search pages
- Product instructions
- Menus
- Recipes

Most consumer texts give information in a short, clear way. Writers of consumer texts know that readers need to get the information quickly and use it right away. Some consumer texts use few words or none at all.

For example, signs use few or no words because you need to read them and act quickly. You see signs every day as you drive, shop, and travel where you live.

railroad crossing

deer crossing

contruction

escalator

no smoking

Most consumer texts use words. Sometimes you will see words you do not know. You can use **context clues** to figure out the meaning of unfamiliar words. Context clues are words, phrases, and sentences that appear near the unfamiliar word. They give hints about the word's meaning.

Example:

Imagine that you are setting up a computer and read an instruction about a USB cable. Imagine that you do not know what a USB cable is. You might use these context clues.

Words or phrases that rename the unfamiliar word:
Find the USB (Universal Serial Bus) cable.
Words or phrases that define the unfamiliar word:
Find the USB cable. A USB cable connects computers to other devices.
Words or phrases that give an example of the unfamiliar word:
A USB cable can be used to connect a device to a power source. For example, many charging cables for cell phones are USB cables.
Images that show what the new word is:
Find the USB cable.

PRACTICE

Draw lines to match some common signs to the phrase that tells what they mean.

1. Smoking is not allowed here.
2. This escalator goes up.
3. Watch for deer crossing the road.
4. Watch for people at work on the road.
5. A railroad crossing is ahead.

a. 

b. 

c. 

d. 

e. 

Read the warning label. Then answer the questions.

WARNING

Keep out of reach of children.
Use in a well-ventilated area. Open windows and doors to create airflow.

 Harmful or fatal if swallowed.

May burn eyes and skin.

FIRST AID

In case of skin or eye contact, flush affected area with cold water for 15 minutes. If swallowed, rinse mouth with cool water. Drink a large glass of water or milk. **Get immediate help** from a doctor or poison control center at **800-555-5555**.

6. What context clue helps you know what "well-ventilated" means?

____

7. How does the image help you understand what "fatal" means?

____

8. What is the label's main purpose—to protect, to describe, or to teach?

____

Read the consumer text about food labels. Then answer the questions.

How to Read a Food Label

First, read the serving size. It tells you how much food you should eat at one time. This label shows that one serving is two tablespoons or 32 grams. All of the facts on the label are for one serving. Note that this jar has more than one serving in it. In fact, it has close to 35 servings in it.

Next, read the facts on calories. Calories are the amount of energy you get when you eat this food. This food has 190 calories.

Then look at the other facts. The label shows how much fat, cholesterol, sodium (salt), carbohydrates, and protein are in the food. Some of them are in grams (g). Some of them are in milligrams (mg). Facts are also given for the % Daily Value (percent of daily value). This percent is based on a 2,000 calorie diet. If you eat 2,000 calories a day, one serving of this food is 25% of your fat calories. That means you should not eat more than four servings of this food in a day.

Last, look at the facts at the bottom. The label will tell you if the food has vitamins or minerals in it. One serving of this food has four percent of the iron you should eat each day. It is not a good source of Vitamin A, Vitamin C, or calcium.

Food label on a jar of peanut butter

9. If someone is trying to eat less cholesterol, should that person eat this peanut butter? Explain how you know.

10. If someone eats four tablespoons of this peanut butter, how many servings has that person eaten?

11. A person uses a serving size of this peanut butter to make a sandwich. How many calories does the serving add to the sandwich?

12. Read the last (bottom) section of the food label. Which of the vitamins or minerals does this peanut butter provide the most of?

LESSON 32: Sequence in Instructions

Instructions teach you how to do something. You have probably read many different types of instructions over the years. Instructions may explain how to do the following things.

- how to put something together
- how to complete a form
- how to make a recipe
- how to operate a machine

All instructions include a set of actions that you need to do. Some also include a list of things you need. The list may include materials, tools, or ingredients.

Some instructions use only text to explain how to do something. Others use only images. Some use both text and images.

Instructions are usually presented in a certain **sequence**, or order of events. If you try to follow the steps out of order, you will not successfully complete the process. Some instructions may use **signal words** to help you follow the steps in the right order. Look for words such as *first, next, then, after that*, and *finally*. Other instructions may use numbered steps. Follow the numbers from 1 to the highest number.

Example:

1. Lather hands with soap

2. Rub both palms together

3. Rub each finger and between fingers

4. Rub palms with finger nails

5. Rub back of hand with finger nails

6. Wash thoroughly and towel dry

This set of instructions is for health care workers. Health care workers must wash their hands well to avoid getting sick. This set of instructions includes text and images. It also uses numbered steps.

PRACTICE

Read this set of instructions. Then answer the questions.

1. Where would you likely find these instructions?

in a building with businesses   in an apartment   in a garage

____

2. The first step is to "pull the nearest fire alarm." Why do you think this is the first step?

____

3. Why does Step 3 stress "DO NOT run"? What might happen if people run?

____

4. Read Step 6 again. Why might someone go back into the building?

____

5. How does Step 5 help the supervisor? What would happen if someone skipped this step?

____

Recipes are instructions. Recipes have a section for ingredients you will need. They also have a section for steps. You must follow the order of steps if you want food to turn out right. Most recipes assume that you know basic information about cooking.

PRACTICE

Read the recipe. Then answer the questions.

Ingredients:

*Grated cheddar cheese, 1 cup
Elbow macaroni, 8-ounce package
Butter, 4 tablespoons
Milk, 2 cups
A little bit of cayenne (red pepper)
Salt and pepper to taste

Instructions:

1. Fill a large pot with about 1½ quarts of water. Bring the water to a boil.
2. Add the macaroni to the pot. Cook on medium heat for 7 to 10 minutes. Turn the heat down if the water starts to boil over.
3. When the macaroni is cooked, drain the water.
4. Put the drained macaroni back into the pot.
5. Stir in butter, milk, grated cheese, and cayenne.
6. Mix until cheese has melted.
7. Add salt and pepper to taste.

**Note: American cheese, Colby cheese, or Swiss cheese can also be used.*

6. It is important to read the whole recipe before you start to cook. Look at the first ingredient on the list. If you have a block of cheese, what would you need to do first?

____

7. Look at the fifth ingredient on the list. What context clue tells readers what cayenne is?

____

8. What common kitchen tool would you need to complete step 3?

____

9. What should you do if the water starts to boil over?

____

LESSON 33: Visuals in Workplace Texts

Workplace texts include instructions, forms, and other documents you read and use for your job. The purpose of most workplace texts is to inform or to explain. Many workplace texts include visuals.

Writers use visuals to present a lot of information in a small amount of space. Writers choose different types of visuals based on their purpose.

Writer's Purpose	Possible Type of Visual
draw attention to specific information	photo, labeled diagram
show how pieces of information relate to one another	graph, chart, or table
show how a process works	flowchart
show important dates	timeline or schedule
show where things are	map

Workplace texts that inform may include these types of documents.

- pay schedules
- maps of an office or delivery route
- information about health insurance
- rules about safety
- rules about use of email and the Internet

Documents that inform often have maps, charts, graphs, tables, or photos. These types of visuals and illustrations help people to interpret writing in the workplace.

Example:

Proud and Company now offers dental plans for workers. If you would like to sign up for a dental plan, please stop by the office. You will need to fill out a form. All forms are due by next Friday.

Dental Plan	Cost per Month
Individual	$10.50
Individual + Spouse	$22.50
Individual + Children	$20.50
Family	$35.50

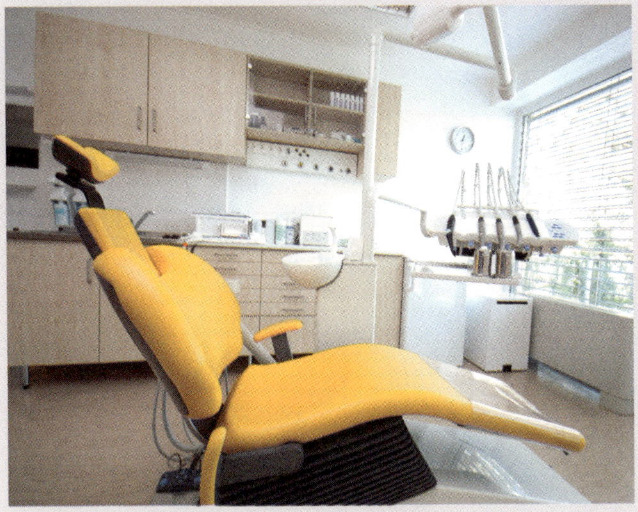

Regular dental visits keep your teeth and gums healthy.

Workplace texts that explain may include these types of documents.

- instructions to ask for days off
- instructions to report sick days
- manuals for operating machines
- instructions to order supplies
- instructions to use office equipment

Documents that explain often have diagrams, flowcharts, or photos.

Example:

Proud and Company Vacation Policy

To request a day off, follow these steps.

Send an email to your supervisor. The subject should be DAY OFF REQUEST. Write the date(s) in the body of the email.

↓

Your supervisor will approve or reject your request. If your request is approved,...

↓

Send an email to SCHEDULE@proudandco.com. Include your name, your supervisor's name, your shift, and the day(s) you will be off.

↓

The scheduling office will check to make sure someone can work your shift. If your request is approved,...

↓

Write your name in ink on the day you will take off on the VACATION calendar in the break room. Place a check next to your name.

PRACTICE

Read the workplace text that informs new workers about tax options. Then answer the questions.

Know Your Tax Options

When starting a new job, you will be asked to fill out a lot of paperwork. It is easy to get overwhelmed. Everyone will need to fill out a W-4. A W-4 is a tax form. Here are a few tips about how tax options can change your take-home pay.

When filling out your tax form, you must decide how many dependents to claim. A dependent is someone who depends on your income for daily survival. For example, if your salary supports your child, that child may be your dependent.

---------- **Separate here and give Form W-4 to your employer. Keep the top part for your records.** ----------

Form **W-4**
Department of the Treasury
Internal Revenue Service

Employee's Withholding Allowance Certificate

▶ Whether you are entitled to claim a certain number of allowances or exemption from withholding is subject to review by the IRS. Your employer may be required to send a copy of this form to the IRS.

OMB No. 1545-0074

1 Your first name and middle initial | Last name | 2 **Your social security number**

Home address (number and street or rural route) | 3 ☐ Single ☐ Married ☐ Married, but withhold at higher Single rate.
Note. If married, but legally separated, or spouse is a nonresident alien, check the "Single" box.

City or town, state, and ZIP code | 4 **If your last name differs from that shown on your social security card, check here. You must call 1-800-772-1213 for a replacement card.** ▶ ☐

5 Total number of allowances you are claiming (from line **H** above **or** from the applicable worksheet on page 2) | **5**

6 Additional amount, if any, you want withheld from each paycheck . . . . | **6** $

7 I claim exemption from withholding for 2015, and I certify that I meet **both** of the following conditions for exemption.
- Last year I had a right to a refund of **all** federal income tax withheld because I had **no** tax liability, **and**
- This year I expect a refund of **all** federal income tax withheld because I expect to have **no** tax liability.

If you meet both conditions, write "Exempt" here . . . . ▶ | **7**

Under penalties of perjury, I declare that I have examined this certificate and, to the best of my knowledge and belief, it is true, correct, and complete.

Employee's signature
(This form is not valid unless you sign it.) ▶ **Date ▶**

8 Employer's name and address (Employer: Complete lines 8 and 10 only if sending to the IRS.) | 9 Office code (optional) | 10 Employer identification number (EIN)

For Privacy Act and Paperwork Reduction Act Notice, see page 2. Cat. No. 10220Q Form **W-4**

If you are single, you should enter either 0 or 1 in Box 5 on the tax form. You can claim yourself as a dependent. The lower your number, the more money will be held back from your paycheck for taxes. This means you will take home less with every paycheck. More money will be withheld throughout the year. You will likely get a tax refund in April.

The higher the number you choose, the less money will be withheld from your paychecks. This means that your paychecks will be higher. But you could owe money for taxes in April. The choice is yours.

1. What important information about paychecks does this text explain?

____

2. Who can be counted as a dependent, according to this information?

____

3. Why do you think the writer included the image of the tax form?

____

____

LESSON 34: Structure in Essays

Essays keep readers informed about important topics and ideas. Essays may give a writer's opinion about a topic. Essays may explain a writer's ideas.

Essays follow a common structure. Understanding an essay's structure helps the reader understand how information is connected. Each part of the essay has its own job.

Introduction The introduction of an essay is often one paragraph. This paragraph is very important. It tells readers the general topic that the essay will explain. It gets the reader's attention. Most importantly, the introduction states the essay's **thesis**. The thesis is the main idea that the writer will talk about. The thesis is often the last sentence in the introduction. It tells the writer's opinion on the topic.

Body The body of an essay has one or more paragraphs. The body supports the thesis of the essay. You may find examples or explanations in the essay's body. You may read reasons that the writer thinks the ideas are important. The body may also present opinions that disagree with the thesis. Then the writer shows why these opinions are not as a strong as the thesis.

Conclusion The conclusion of an essay is usually one paragraph. The conclusion ends the essay. It reminds readers of the thesis and main points. It connects all the ideas that were explained earlier. The conclusion may also ask readers to take action or to think more about the topic.

You will find essays in newspapers and magazines, online, and in books. Many blogs are informal online essays. The blogger wants to make a point—that is, state a thesis. The blogger hopes that readers will think about the thesis long after they finish reading the blog.

PRACTICE

Here is part of a blog from a blogger who writes about workplace issues. This blog entry is about equal pay for men and women.

Read the introduction to the blog. Then answer the questions.

WORKPLACE VOICE BLOG

HOME ABOUT BLOGS CONTACT

Equal Pay for All of Our Daughters and Sons *by Jacob Smith*

When I was single, I did not think about equal pay for men and women. The issue did not hurt me, so I ignored it. But now I am married and the father of two sons and a daughter. They are still young, but their days of work are not too far off. I tell all three kids to do well in school. I tell them to think about what they like to do. I hope that all three find work they love some day. I want opportunities for my daughter, just as I do for my sons. Suddenly, I have become a supporter of equal pay. Our daughters and sons should get the same pay if they do the same work.

1. What is the general topic of the blog entry?

2. Why is the blogger interested in the topic?

3. Write the thesis, or main idea, of the introduction.

Now read the body of the blog entry. Remember that the body paragraphs develop the thesis, or main idea. Then answer the questions.

The law says that men and women who do the same work must be paid the same. But in fact, women are often paid less than men. Why is this? In the past, men and women often held different types of jobs. Women had limited work options. They often worked as caregivers, teachers, or helpers. Men had many more options. They could work in law, business, industry, medicine, and many other fields. The work that men did paid more. So the idea that "women's jobs" should pay less than "men's jobs" made sense to many people.

But today, men and women work in the same kinds of jobs. They do the same tasks. They must have the same skills to succeed. Today, many people say that pay should be equal because the work is equal. They suggest that companies should look at the job duties and rate jobs. Pay would be based on those job ratings. If pay is based on the job ratings, then companies can set a fair wage. It would not matter who was doing the job.

Other people disagree. They say that the laws of supply and demand make wages fair. These people think that rating job duties is bad for supply and demand. It keeps businesses from hiring and paying as they want to.

4. According to the first body paragraph, why did some people assume that women's jobs should pay less than men's jobs?

5. The second and third body paragraphs explain two different views on equal pay. What are the two different views?

Read the blog's conclusion. Then answer the questions.

When I think about my kids' futures, I find that I agree with the idea of rating jobs. Say that my son and my daughter both decide to become nurses. Both do the same kinds of work. Both work long hours in a hospital. Both need the same skills, and both face the same kinds of problems. Should my son get paid more just because he's a man? Of course not! The same work deserves the same pay. From now on, I'll pay more attention to this issue. I'll do what I can to be sure that the law is followed. If you have children, I encourage you to do so, too. Our kids' futures depend on it!

Male nurses and female nurses should get equal pay for the same work.

6. Write the restatement of the thesis, or main idea, in this conclusion. You may want to look back at the introduction and review the thesis there.

____

7. How does imagining the future help the blogger decide what he thinks about equal pay issues?

____

____

8. The blogger ends the entry by telling what he plans to do about the issue. What does he ask readers to do?

____

____

LESSON 35: Compare Information in Ads

Ads come in many forms. You see them in many places. They may be as large as a billboard. They may be as small as a pop-up on your phone. They may use only words. They may have a mix of words and images. When ads have images or other supporting illustrations, you should pay attention to them. The images may be there simply to provide information. They may also be there to make us feel a certain way.

Whenever you see ads, you need to understand their messages. You can get useful information from ads. But you must be able to filter out information that you do not need. You must also watch out for misleading information. Then you can compare products and services and choose the best one for your needs.

Conflicting Purposes

When you read an ad, you are looking for information. The business that creates the ad has a different purpose, or reason for writing.

Your Purpose	The Business's Purpose
To get information about what to buy	To persuade you to choose this product and not a competitor's product

Do not get distracted by the business's purpose. Look for information. Do not let pretty images, bright colors, or upbeat music distract you from the information you need.

PRACTICE

Read and answer questions about three ads for milk. Then answer the questions to compare the ads.

To Your Health!

Why milk?

Because it's the whole package—from head to toe, your body will be glad you drink milk!

- Silky hair . . . glowing skin . . . milk helps you look good!
- Stronger bones . . . faster muscle recovery after exercise . . . milk helps you stay fit!
- Calmer tummy . . . better sleep . . . milk helps you manage life's little stresses!

Milk—for your health and the health of your family!

1. Ads are created to appeal to specific groups of people. What is the target group of this ad? (Hint: Look at the ad's title, "To Your Health!")

____

2. Which fact in the ad might appeal to an athlete?

____

We Do Dairy the Old-Fashioned Way

Love milk? So do we! At Ollie's Old-Fashioned Farms, we know each of our dairy cows by name. Our family has managed the dairy for three generations. And while we may not use a milking stool and pail anymore, we still respect old-fashioned values like quality, customer service, and family!

Our cows graze freely. They are never confined. We use the gentlest milking methods. We guarantee the health of our herd.

Small-batch dairy products—that's what we're all about! We offer

- the freshest whole and skim milk
- cream richer than what you can buy at the grocery store
- handmade cheeses and yogurts

Come visit our dairy! Show your kids where their favorite dairy treats are made. Then stop by our market for the best dairy you've ever tasted!

Underline the answer that best completes the sentence.

3. The image in the ad shows ____________________.

   cows grazing freely   how farmers long ago milked cows

4. The last paragraph of the ad suggests that the target audience is ____________________.

   parents   dairy farmers   children

Answer the question.

5. The ad says that the people at Old-Fashioned Farms "respect old-fashioned values like quality, customer service, and family!" Which of these values does not give readers information about the dairy's products? Why do you think the ad includes this value?

   ____

   ____

   ____

   ____

Worried About Your Milk?

Try Ours Instead!

If you're like a lot of parents, you keep milk in your fridge. It's a healthy choice for growing kids. But these days, it's hard to know what might be in your milk.

Growth hormones? Antibiotics?

MoreMoo Milk guarantees that all our dairy products are free from the harmful chemicals that are in many brands today. How can we make this promise?

- Our farmers feed their cows the right food in the right amounts. Healthy cows produce more and better milk!
- Our farmers never use artificial growth hormones. They do not use antibiotics.
- We test every batch of milk to be sure it's pure. We will never put milk with traces of growth hormones in our products.
- We deliver our milk only to stores within a day's driving distance. That keeps it fresh, too!

Try **MoreMoo Milk**! Pour that glass full—know that you're serving your family pure, fresh goodness!

Answer the questions.

6. Ads sometimes address concerns or fears that people have. What concerns does this ad address?

____

____

7. Study the image in the ad. What idea do you think the ad tries to communicate?

____

____

Now compare the ads and underline the correct answer or answers for each question.

8. Which ad includes the most facts that you could check online about milk?

To Your Health! We Do Dairy the Old-Fashioned Way

9. Which ads appeals to readers' worries about modern dairy farming?

To Your Health! We Do Dairy the Old-Fashioned Way Worried About Your Milk?

10. Which ad relies most on readers' positive feelings about family and tradition?

We Do Dairy the Old-Fashioned Way Worried About Your Milk?

Circle the correct answer.

1. What information do writers include in a biography?
   A. information about a person's life
   B. information about steps in a process
   C. information that sums up an argument
   D. information that compares two products

2. Which signal words help readers keep track of details arranged in time order?
   A. also, and, in the same way
   B. in short, therefore, to sum up
   C. early on, soon afterward, in the end
   D. because, for this purpose, as a result

3. What is the main purpose of an essay's introduction?
   A. to list examples
   B. to develop ideas fully
   C. to state the writer's thesis
   D. to sum up the writer's argument

4. What is the main purpose of an essay's conclusion?
   A. to present opposing ideas
   B. to provide support for the main point
   C. to remind readers of the essay's main point
   D. to get readers interested in the essay's topic

5. What is the main purpose of an essay?
   A. to teach readers to use a process
   B. to get readers to think about a topic
   C. to persuade readers to buy a product
   D. to inform readers about their tasks at work

Read the consumer text. Then circle the correct answer.

Office Memo: Spam
To: All Employees
Subject: Technical Support Team

Spam is unwanted email. Spam is a frustrating time-waster. Even worse, spam puts our business at risk. Hackers are people who try to break into computer systems. They often use spam to do this.

Starting today, all employees must follow these steps to cut down on spam messages.

1. Never reply to spam. Some email senders promise to delete your address from their mailing lists if you click on a certain link in the email. However, if you click the link, you tell spammers that they have reached a working email address. Then they are very likely to sell your email address to other spammers.

2. Turn on your spam filters. The company's email system has good spam filters. These filters help keep spam out of your email box. Enable these filters. Once you turn on these filters, most spam will go directly into the filter folders. Then you can review the emails in those folders. If they are spam, you can delete them.

Not sure how to set up your spam folder? Contact our Technical Support team.

6. What is the purpose of the memo?
   A. It tells how to write spam.
   B. It gives examples of why spam is risky.
   C. It explains how to protect the company from spam.
   D. It gives the steps to set up a spam folder.

7. Read the sentences from Step 2 of the memo.

   Enable these filters. Once you turn on these filters, most spam will go directly into the filter folders.

   Which words provide a context clue that helps you guess what *Enable* means?
   A. turn on
   B. most spam
   C. go directly
   D. filter folders

8. When does a spam filter work?

   A. before an employee receives the spam

   B. before an employee reads the spam

   C. after an employee reads the spam

   D. after an employee replies to the spam

9. According to Step 1, how do spammers trick employees?

   A. by sending them several spam messages

   B. by agreeing to delete their email addresses

   C. by sending their messages to the spam filter

   D. by selling employees' email addresses

10. How does the Technical Support team help employees who read this memo?

   A. It explains how spammers get employees' emails.

   B. It explains how the business might be harmed by spam.

   C. It explains how employees can get more email training.

   D. It explains how to keep spam out of the employee's inbox.

11. If employees need more help setting up the spam filter, who can they contact?

   A. a manager

   B. a spammer

   C. an employee in Technology Support

   D. an employee who knows how

Read and compare these ads for places to live. Then circle the correct answer.

The Gables

1 mile from city center, short walk to stores and offices
$700 per month
1 bdr

Highlights:
- 2 references required
- FREE cable and water
- Large bathroom with bathtub and shower

Northgate Apartments

Downtown location
$900 per month
1 bdr

Highlights:
- References required
- Unfurnished
- Ceiling fans
- Deposit required for pets

House for rent

North District
$950 per month with 12-month lease
2 bdr, 1.5 bath

Highlights:
- All new appliances • Schools within walking distance
- Nice covered patio • Attached carport • Deposit required for cats

Garage apartment to rent

College District
Will negotiate rent
1 bdr

Highlights:
- Small appliances • Parking on site • Walking distance to college

12. Chandra is looking for an apartment. She is a single professional and needs a small apartment. She likes to walk and watch TV. Which ad best describes her needs?

A. the ad for The Gables

B. the ad for Northgate Apartments

C. the ad for the house

D. the ad for the garage apartment

13. Theo is a student who can't afford a high rent. He is working his way through college. Which ad best describes his needs?

A. the ad for The Gables

B. the ad for Northgate Apartments

C. the ad for the house

D. the ad for the garage apartment

14. Which ad would *not* appeal to a renter with two cats?

A. the ad for The Gables

B. the ad for Northgate Apartments

C. the ad for the house

D. the ad for the garage apartment

15. Which ad describes a place suitable for a family with young kids?

A. the ad for The Gables

B. the ad for Northgate Apartments

C. the ad for the house

D. the ad for the garage apartment

LESSON 36: Setting and Character

Setting is when and where a story takes place. Stories, plays, movies, and even some poems have settings.

Stories can use many kinds of settings.

Realistic settings are like the real world. Realistic settings might be like today's world, or they might be like a time in the past. The setting could be in the summertime at the local park. Or the setting might be in Boston during the American Revolution. But the things you find in a realistic setting actually exist or existed in the past. They are, or were, real.

Fantastic settings may have some real-world parts. But they also have pretend parts. For example, you may tell your children a fairy tale. It has a castle and a prince. These things could really exist. But the fairy tale also has a magic wand, a horse that can fly, and a mirror that talks. These things do not really exist. They are fantasy, or made up ideas.

Futuristic settings show what an author imagines the world might be like in the future. You often see futuristic settings in science fiction stories and movies. They might feature cars that float, houses that clean themselves, and people that live to be three hundred years old. These things do not exist now. Someday, the author thinks, they might.

When you study a story's setting, ask these questions.

- Where is the story set? Is it in a city, or in a village, or on another planet? Is it a real place?
- When is the story set? It can be set in the past, in the present, or in the future. It can be set in a pretend time, too, like a tale that begins with "Long ago" or "Once upon a time."
- What kinds of people live in this time and place? Are they real or made up people?
- What kinds of things do the people use in this time and place? Just like in a movie or a play, all the objects in a story are part of its setting.

When you understand a story's setting, you will better understand what the characters do. Characters act and speak in a story. **Characters** are usually people, but not always. Animals and objects act and speak in some stories.

When you read a story, you get to know the characters, just like you get to know people in real life. By looking at these details about the characters, you can better understand them.

- You know what they do and say.
- You often know what they think.
- You know what other characters say about them.
- You know how they dress, where they live, and what they own.

These details help you figure out what kind of people the characters are. Then you can compare the characters to people in your own life. You can figure out why they do things.

Fiction has real and made up parts. When you read fiction, decide who the main characters are. These are the characters that the story is mostly about. Get to know them best. You will meet minor characters, too. These characters interact with the main characters but are not as important. Notice how minor characters change the main characters'experiences.

PRACTICE

Read the story. Then answer the questions about setting and character.

The Steadfast Tin Soldier

by Hans Christian Andersen

Long ago, a toymaker made twenty-five tin soldiers from the same old tin spoon. But one soldier had only one leg because there wasn't enough metal left to finish him. This soldier, who stood well on his one leg, was in love with a little maiden. She was a paper doll. She was a dancer with one leg stretched high in the air. She stood on one leg just like him. She wore a shiny charm on a pretty ribbon around her neck.

One morning the children who owned these toys put the tin soldier on a windowsill. Suddenly, he fell out the window. He landed far below. The children hurried outside to look for him. They didn't see him. So they went back inside. The scared little soldier could have shouted, "Here I am!" But he did not think it proper to shout when he was in uniform.

Later, two other boys came along and spotted him. It had begun to rain, so they made him a boat from folded newspaper and sent him sailing down the gutter. The boat soon entered a long, dark tunnel. The soldier was scared and wished the maiden were with him. Then he heard a loud roar, and his boat shot down a waterfall into a huge waterway. It soon sank.

Just as the soldier tumbled out of the paper boat and into the water, a big fish came by and swallowed him. It was so dark inside that fish! Then the fish began rushing about. Suddenly the fish was still, and the soldier saw daylight. Someone had caught the fish and sold it at the market.

A cook bought the fish and took it to his kitchen. As he cut the fish open, he cried, "A tin soldier!" The cook took the soldier to the living room to show the toy to the family. Wonder of wonders! The soldier saw that he had come home! Around him were the same children—and the little maiden! He looked at her, and she looked at him, but they never said a word.

Then, for no good reason, one of the children picked up the tin soldier and threw him into the fireplace. Just as quickly, a puff of wind blew the paper maiden into the fire, too! The next morning, all that was left in the ashes were the maiden's charm. It was burnt black as coal. Next to it was a little lump of tin, melted into the shape of a heart.

1. Is the story set in the past, in the present, or in the future? ______________________

2. What details in the story tell you when it is set?

____

3. Does the story have a realistic, fantastic, or futuristic setting? ______________________

4. What details in the story help you decide what kind of setting it has?

____

5. Who are the main characters in the story? List the two most important characters.

____

6. Who are the minor characters in the story? List the minor characters.

____

7. What does the tin soldier want most? Use a detail from the story to explain how you know.

____

8. Think about how the story ends. Does the tin soldier get what he wants most? Explain.

____

____

LESSON 37: Plot and Resolution

Plot is what happens in a story. It is the action and the order in which the action takes place. You might ask a friend who has seen a movie, "What happened in the movie?" You are asking about the plot.

Many plots have four building blocks.

The *basic situation* is the first part of the plot. It answers important questions about the story.

- Who are the main characters?
- What is their basic problem? That problem is the story's conflict. Conflict happens when a character wants or needs something but, for some reason, can't get it. There can be conflict with the situation. Or there can be conflict within the character's own mind and emotions. There can be many conflicts in the story.

Complications make up the second part of the plot. In this part of the story, many events happen. Problems come up. Readers wonder how the main character will solve the problems. Readers feel suspense, or anxiety. They don't know what the characters will do.

Climax is the third part of the plot. The climax is the most exciting moment in the story. By the time you read the climax, you really want to know what is going to happen. How will the characters solve the conflict? The climax is when the character must make an important choice to solve a problem.

The *resolution* is the final part of the plot. The characters solve their problems and the story ends. Readers no longer feel suspense. Instead, they feel a sense of completion.

Readers can follow the parts of the plot in a diagram like this.

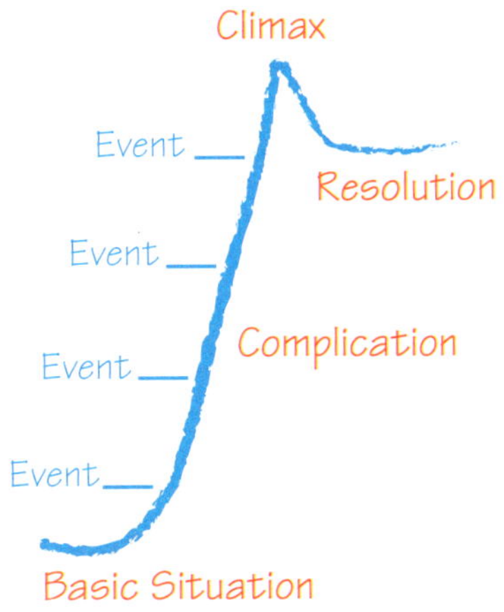

Some stories end happily. The characters solve their problems and get what they want or need. Other stories end sadly. The problems are too hard for characters to solve. Often, characters in a sad story learn something or change. Even though the ending is sad, the story still has a resolution.

PRACTICE

Read the story. Then answer the questions.

Music to Her Ears

Each morning when Anna walked across the yard to get the mail, she heard the music. Her neighbor, Mrs. Marsh, practiced piano every morning. In pleasant weather, the music flowed through the open window. Anna could hear it clearly. On cold days, she still heard the soft sounds. And each morning, Anna felt a bit sad when she heard the music.

Before she moved from the city, Anna had played the piano nearly every day. She hadn't owned a piano. With all the bills, not much money was left for hobbies. But she had lived near a small church where a friend was a pianist. Anna was welcome to play the church's piano on weekdays. She wanted to keep up the skills she learned during childhood lessons. She played the piano in the church every Monday. She loved to practice.

Anna was happy in the house she and her family rented away from the city. The kids could play in the lawn and walk to their school a few blocks away. Still, when she heard Mrs. Marsh practice, Anna felt homesick for the city. She missed "her" piano.

One morning, Anna noticed Mrs. Marsh sitting on the steps next to the mailboxes. The older woman seemed as if she did not feel well. So Anna went out to see if she was okay.

"Oh, thank you, dear," Mrs. Marsh said as Anna helped her back to the house. "Sometimes I feel weak, and I have to sit down—or I might fall down!"

Mrs. Marsh sighed. "Growing older comes with challenges, I find. It's getting harder for me to do chores, especially in the yard. I may have to hire someone to help, but I really can't afford it. A retired piano teacher usually doesn't put a fortune away for her later years!"

Anna told Mrs. Marsh to call her if she felt weak again. She left the older woman her phone number and walked back to her house. Just before Anna shut her front door, she heard the piano music again. Her eyes lit up with a sudden thought, and she hurried back to Mrs. Marsh's house.

"May I suggest a trade?" Anna asked. She explained how much she missed playing the piano. "Maybe I can help you around the house, with the harder chores. And you could allow me to practice on your piano."

Mrs. Marsh was thrilled to have a neighbor who shared her love of music. She agreed to the idea right away. The next day, when Anna came to practice, she found a surprise on the piano's music holder. "A duet!" she exclaimed. She recognized the music, which was written for two people to play at once.

Mrs. Marsh nodded. "After you've had some time to play, I hope that you might like to join me at the keyboard."

Anna touched the well-worn keys and smiled at the sweet sound coming, this time, from her own hands.

1. Read the details from the story. Circle the details that are part of the basic situation.

   Anna has moved to a house away from the city.

   Anna took piano lessons when she was young.

   Anna does not have a piano to play.

   Anna feels sad when she hears Mrs. Marsh play.

   Anna helps Mrs. Marsh when she is weak.

   Anna offers to exchange chores for time at the piano

2. Read the events that make up the plot's complications. Then draw a line to the plot diagram to show the order in which they happen. Remember: the early events are low on the diagram. They build to the climax, the high point of the story.

   Mrs. Marsh says that she is a retired piano teacher.

   Mrs. Marsh feels weak and sits by the mailboxes.

   Anna tells Mrs. Marsh to call if she feels weak again.

   Mrs. Marsh explains that she can no longer do all her chores.

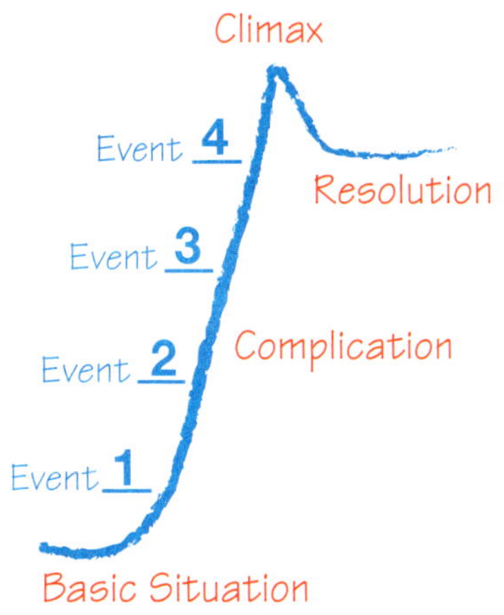

3. What happens at the plot's climax? Retell the event in your own words.

   ____

   ____

   ____

   ____

4. Describe the story's resolution. Does the story have a happy ending? Does Anna solve her problem or get what she wants?

   ____

   ____

   ____

   ____

LESSON 38: Sequence of Events

Sequence means time order. When things happen in a story, they happen in a sequence. That means most stories are arranged in time order. Stories have a beginning, middle, and end. Something happens first. Then other things happen in the middle. Then something happens last. The events are arranged in the order in which they happened.

Sequence of events refers to the order in which events take place. Keeping track of the sequence of events helps you in two ways.

- You can better understand what happens in the story.
- You can recall details more easily.

To find the sequence in a story, look for sequence clue words and time-order signal words.

Sequence Clue Words	
times of day	morning, afternoon, evening 6:00 p.m., noon, sunset
days of the week	Monday, Tuesday, Wednesday, Thursday, Friday, Saturday, Sunday
months	January, February, March, April, May, June, July, August, September, October, November, December
time of year	winter, spring, summer, fall
years	1642, 1776, 1820, 1965, 2015

Time-Order Signal Words		
after	first	since
after a while	following that	since then
at last	in the end	soon
at the beginning	last	soon after
before	later on	then
earlier	long ago	third
early on	next	to start with
finally	second	when

A sequence chart can help you keep track of the order of events when you read a story.

PRACTICE

Read the folktale, or legend, from Indonesia. Answer the questions about the sequence of events.

People in Indonesia tell stories about Cicak and Kancil. Cicak is a gecko, a type of lizard. Kancil looks like a very small deer. Cicak and Kancil do not get along with each other.

1 One morning, Cicak and Kancil fought about who could run faster. Kancil dared Cicak to a race. They would race from Kancil's house to a big mango tree and back to the house. The one who won the race could do whatever he wanted to the one who lost. Cicak knew that Kancil could run faster, but Cicak agreed to the race anyway.

2 That night, Cicak thought hard about how he could win the race. He had an idea. He begged two geckos to help him win the race.

3 The next day as the race began, Kancil got way ahead. Then he stopped so a tourist could take his picture. Kancil thought he saw Cicak run around the corner ahead of him. He didn't know that this gecko was not Cicak. After he saw the gecko, Kancil ran even faster. Just as he reached the mango tree, he thought he saw Cicak going back to the house. He didn't know that this gecko was also Cicak's friend. Kancil ran as fast as he could back to his house. When he got there, Cicak was already leaning against the house. "I beat you," Cicak said with a big smile. Kancil fell to the ground, exhausted. "Okay, you won," he panted. "What are you going to do to me?"

4 Cicak decided not to be too mean to Kancil. After all, Kancil might one day learn that Cicak tricked him. Then what might happen? So Cicak said, "If everyone finds out how fast we geckos can run, we will have to race all the other animals. So you must not tell anyone about our race. Also, I want you to let us geckos live in your house and catch the insects there."

5 Kancil smiled. That wouldn't be too hard at all! And ever since that time, geckos have lived inside houses in Indonesia.

1. What sequence clue words are used in paragraphs 1 and 2?

2. What time-order signal words are used in paragraph 3?

3. Write the events from the middle of the story in the sequence chart.

Kancil runs quickly to the mango tree.
Kancil sees a gecko going back to the house.
The race begins, and Kancil gets ahead.
Kancil sees a gecko run around the corner.
Kancil runs to his house, but Cicak is already there.
Kancil stops to take a photo.

Event 1 → **Event 2** → **Event 3** →

Event 4 → **Event 5** → **Event 6**

LESSON 39: Make Inferences and Draw Conclusions

To **infer** is to connect something you know and something you read. Once you have made the connection, you can come up with a new idea based on the connection.

What you already know + New information you read = Inference

To draw conclusions, you add up all the inferences you have made and make a judgment.
Inference + Inference = Conclusion

When you read works of fiction or watch a movie, you make inferences and draw conclusions so you can understand the story.

- You infer what motivates people and characters in stories. You connect what they do and say to your real-world experiences.
- You infer what a new word means by connecting it to what you understand.
- You infer themes—big ideas the story points to—from the whole work.
- You infer to connect how characters, events, and themes fit with or challenge what you know about the world around you.

When you connect all of these inferences, you draw a conclusion about what it means for the character and overall story.

Think about the well-known story of Cinderella. You may have shared this story with your children. But neither you nor your children live in a fantasy land where magical things happen. How are you able to understand the story?

You make inferences and draw conclusions.

You read about Cinderella doing chores, putting up with her mean family, and smiling even though she has hard days. So you infer that Cinderella has a good heart and a kind nature. She's strong and hard-working.

Then you read about a ball. You may not know what a ball is, but the characters talk happily about fancy dresses, dancing, and music. So you infer that a ball is a fancy party where people dance.

You read that Cinderella dances with the prince. She ends up marrying him against all odds. She was once poor and badly treated. She is now rich and happy. You infer that she earned her "happy ending" by acting with kindness and courage.

Then you think about that theme, or main message, of the story. Are brave, kind people always rewarded in the real world for their actions? You might think of times when you have seen that happen. You might think of examples that don't fit that theme. You can use all this information to draw a conclusion about the story and the real world.

Stories are part of every culture because readers enjoy seeing what characters do and what happens to them. Readers are interested in how stories and themes match their own experiences or challenge their experiences. Readers enjoy stories because they know how to make inferences and draw conclusions.

Your knowledge and experience in the real world + What happens in the story's world = Inference

PRACTICE

Read an old story from Korea. As you read, think about the characters. What motivates their words, actions, and thoughts? Is their experience like or unlike yours? Then answer the questions.

The Looking Glass

Long ago, a young man traveled to Seoul, Korea. A bright object caught his attention in a store. When he looked at it, he could see his face! He had seen his face reflected in a sink before but never so clearly.

He bought the looking glass. He took it home. He hid it in a drawer. His wife, however, saw him peeking in the drawer and smiling. One day, while he was away from the house, his wife opened the drawer. She found the looking glass. She was startled by the sudden appearance of a very attractive but puzzled young woman! Jealous, she rushed to her mother-in-law. She showed her the glass. "My husband came home from Seoul with a mistress!" she cried.

But when her mother-in-law looked into the glass, all she saw was a wrinkled old woman. "Dear," she said, "all I see is a granny visiting from another village."

Then her father-in-law looked, but all he saw was an old man with gray hair. Thinking he saw his deceased father, he quickly kneeled down and bowed.

The daughter-in-law grabbed the mirror and looked again. There was that pretty woman again. She yelled at the woman. The woman yelled back! Angrily, she threw the looking glass on the floor. It broke into pieces.

1. Read the first paragraph again. The young man is amazed to see his reflection in the looking glass. What can you infer about the young man?

____

____

____

2. Think about how the wife, mother-in-law, and father-in-law respond to what they see in the looking glass. Think about what you know about mirrors. What inference can you make about all three of the people in the story?

____

____

____

3. What can you conclude about why the young man hides the looking glass rather than sharing it with his wife?

____

____

____

4. Why does the wife think that her husband has a mistress? (Remember that she sees him smile when he looks at the looking glass.)

5. What can you infer about whether the wife trusts her husband?

6. When the father-in-law looks into the glass, he bows because he thinks he sees his father. What conclusion can you draw about fathers and sons in Korea at the time of this story?

7. When the young man comes home and finds the marvelous looking glass broken, how do you think he will respond? What details lead you to draw this conclusion?

8. One theme of this story is that people do not always understand what they see. Draw a conclusion about this theme. Does it match your experience, or does it challenge your experience?

LESSON 40: Identify Themes

The **theme** of a work of literature is a big idea, or message, that the author hopes readers will think about. A theme might be one of the following.

- a lesson readers can learn about life
- an idea about human experience that the author wants readers to think about
- a discovery readers make about human relationships

Theme and plot are not the same. A plot is what happens in a story—the events and how they end. Theme is the big idea that grows out of the plot.

Works of literature usually have at least one theme. They often have more than one theme.

Most authors do not state themes in their writing. Instead, readers must discover the theme. Here are steps you can use to identify theme.

- Describe how main characters change through the story. What do these changes say about human experience?
- Think about important moments in the story. What ideas about life do these show?
- Pay attention to symbols in the story—objects that seemed to be linked to special meanings or ideas. Why does the author use the symbol? How do these symbols connect to the theme?
- Connect your answers to these questions with your own experiences of life. Does the story's theme match what you think? Or does it conflict with what you think?

Theme isn't a "right or wrong" idea. You may agree or disagree with the author. What matters is that you think about the big ideas of the works of fiction you read.

PRACTICE

Read the fable made famous by Aesop. Then answer the questions about theme.

The Bear and the Two Travelers

Two men were traveling through a thick forest. Suddenly, a huge bear appeared in their path. One man quickly climbed a tree and hid in the branches. The other man fell to the ground. When the bear came close to him, the man held his breath. He pretended to be dead.

The bear sniffed the man on ground, especially his head. Then the bear turned and continued through the woods. When the bear had gone, the first man climbed down from the tree.

"Tell me what the bear whispered in your ear," he said to the other man.

The other man studied him for a minute. Then he said, "The bear gave me some advice. He said not to travel with a friend who leaves you alone when danger approaches."

An unhappy or painful event often tests the sincerity of friendship.

1. In fables, the theme is often stated. Where is the theme stated in this fable? What lesson does the fable's author hope readers learn about human experience?

____

2. Think about the two travelers. How has their relationship changed by the end of the story? What happened to change it?

____

____

Now read another fable also made famous by Aesop. Then answer the questions.

The Four Oxen and the Lion

A lion often hunted in a field where four oxen lived. The lion had tried to attack the oxen. The oxen stood in a circle. They stood with their tails in the middle every time the lion approached. The oxen were able to fight off the lion. They used their sharp horns to keep the lion away. No matter how or where the lion charged them, they stood together.

One day, the oxen had a disagreement. Each ox said that he was the strongest and bravest. Each wanted all the credit for chasing the lion away.

Finally, the oxen became so angry that they went their own ways. Each went off to eat grass alone. The lion was listening to the argument. He was waiting for the right time to act. When the oxen separated, the lion attacked them, one by one. Soon there were no oxen left in the field.

3. Describe how the oxen act at the beginning of the fable.

4. What causes the oxen to argue?

5. Describe how the oxen act after the argument.

6. What happens to the oxen because of their argument?

7. What is the theme of the story? What lesson about life might the fable's author want readers to think about?

LESSON 41: Idioms and Collocations

An **idiom** is an expression with a meaning different from its literal meaning. At the end of a long day, a friend might say, "Time to hit the hay." The friend doesn't intend to punch a stack of horse food. He means that he is going to bed. This idiom may have come from a time when some people slept on piles of hay in barns. It doesn't mean that today.

Every language has idioms. Since their literal meaning is not what speakers intend to mean, you have to learn them one by one. Some make sense when you think about them. Others don't.

- *To make hay* doesn't mean "to harvest food for cows." It means "to earn money."
- *Make hay while the sun shines* isn't advice for farmers. It means "take advantage of a chance to do something—before that chance passes."
- A "haymaker" is a powerful hit in boxing.

When you read an unfamiliar idiom, you can look it up online or in a dictionary.

PRACTICE

Read the sentences. Each idiom is underlined. Write what the idiom means in the sentence. Check the meaning online if you need to.

1. The two boys argued often because they could hardly ever see eye to eye.

   ____

2. Nick was biding his time until it was his turn to speak.

   ____

3. You took quite a spill from your bike.

   ____

4. One day, Rip's wife was harping on all his undone work.

   ____

5. The woman was sad that her mother had passed away not too long ago.

   ____

A **collocation** is two or more words that are often used together in English. Think of collocations as words that go together. These word combinations just sound right when you use them.

Verbs, or action words, that get a lot of use are often found in collocations. Over time, that use has paired the words. The longer you study a language, the more natural collocations will sound.

Read these common collocations. The verb and the word it is paired with are underlined.

Examples:

<u>Take</u> a <u>break</u>. (Stop work for a short rest.)

<u>Make</u> <u>time</u> for something. (Set aside time for a particular reason.)

<u>Do</u> a good <u>job</u>. (Work well.)

<u>Keep</u> a <u>secret</u>. (Tell no one.)

<u>Go</u> <u>missing</u>. (Become lost.)

<u>Get</u> <u>ready</u>. (Prepare yourself.)

<u>Come</u> <u>in</u> first. (Win, as in a competition or race.)

<u>Break</u> the <u>news</u>. (Tell someone information that is often not good.)

PRACTICE

Use words from the box to write common collocations that use the verbs given. An example is done for you. Check a dictionary, if needed.

notes	trouble	again	the bed
away	calm	busy	crazy
your best	a habit	a chance	married
even	a favor	back	an appointment

6. Take **a chance**.
Take ______________________.

7. Go ______________________.
Go ______________________.

8. Make ______________________.
Make ______________________.

9. Do ______________________.
Do ______________________.

10. Keep ______________________.
Keep ______________________.

11. Break ______________________.
Break ______________________.

12. Come ______________________.
Come ______________________.

13. Get ______________________.
Get ______________________.

LESSON 42: Responding to a Work of Literature

Every story, play, or poem you read has many elements or parts.

- plot and resolution
- setting
- characters
- themes
- interesting word choices

Every time to you read a work of literature, you put reading skills to use.

- making inferences
- drawing conclusions
- figuring out idioms
- connecting the story to your experience

PRACTICE

Read the story. Then answer the questions.

Stitches

Maris frowned as she sat on the exam table in the doctor's office. "No stitches!" she said. "They'll hurt!"

Papa hugged Maris. "They'll hurt a little," he agreed. "But you need them to heal right."

Dr. Grant nodded. "You took quite a spill from your bike. The cut on your arm won't heal quickly without stitches."

Maris hid her face against her father's chest. "I don't like needles," she insisted. "I don't like shots."

"Just one little shot," Dr. Grant assured her. "Then you won't feel the stitches at all."

"Maris," Papa said. "Do you remember hearing about your great-great-grandmother?"

"The one I'm named for?" Maris sniffed. "Who lived in Europe?"

"Yes, that's her," Papa said. "She was clever. Once, she even figured out how to stitch up the sky."

"Oh, Papa, don't be silly," Maris giggled.

"It's true!" Papa said.

Maris had been the village's tailor. She was so skilled with needle and thread that the villagers said she could fix anything. She stitched soles back onto torn boots. She stitched babies' blankets. She made clothes for everyone around. There wasn't a house in the village that didn't have one of her fine quilts on the bed. Stitching was Maris's special talent.

One day, a violent thunderstorm blew through. The villagers huddled inside and waited for it to pass. Loud cracks and booms of lightning tore at the sky. And when the storm was gone, the villagers found that the sky had in fact been torn open. A large flap of sky was hanging down toward Earth. Through the tear, the villagers could see stardust falling.

Now, you might think that this was a pretty sight. But stars are fire! The stardust that fell down to Earth started a little fire here and a little fire there. The villagers rushed about with pails of water, putting out one fire after another. First a wheat crop would start to burn. Then a house's roof would spark. The villagers were soon bone tired!

"Someone must fix the tear in the sky," the village mayor said. The villagers thought and thought. No one knew how to solve the problem. Then Maris spoke up.

"I have some special thread," she said. "I've been saving it for a special time. Now may be that time. It's the strongest thread I know. It is as strong as iron! It can stitch up the sky."

"But how will we get the thread to the sky?" the mayor asked.

Just then, the best archer in the village stepped up. "My arrow can be the needle. My strong arm can shoot the thread to the sky."

Maris and the archer stood beneath the torn flap of sky. They tied the iron thread to the arrow. The archer took careful aim. Zwing! Up flew his arrow. It just caught the torn edge of the sky. When it fell again, it pulled the thread behind it.

"Again!" cried Maris. Children ran to bring the arrow back. The archer shot it again. And again! Maris pulled the thread tight each time. Slowly, the thread pulled the pieces of sky together again. No more stardust could fall through. The village was saved! Maris and the archer were treated to a feast!

"All done," said Dr. Grant. She covered the stitches with gauze and tape.

"That's it?" Maris asked. "That wasn't so bad."

"Of course not!" Papa agreed. "You and Grandmother Maris both know your way around stitches!"

1. Why does Maris's father tell her a story about her great-great-grandmother?

____

____

____

2. Dr. Grant says that Maris "took quite a spill" from her bike. What does *spill* mean in this sentence?

____

____

____

3. What problem do the villagers face in Papa's story?

____

____

____

4. Read the events that make up the plot. Then draw a line to the plot diagram to show the order in which they happen.

Maris thinks of a way to fix the sky.

Stardust falls and starts little fires.

An archer helps Maris carry out her plan.

A storm tears a hole in the sky.

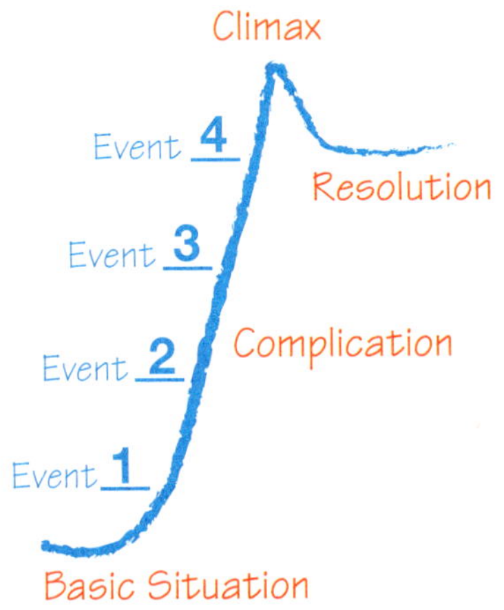

5. What happens during the resolution of the story of Maris and the villagers?

6. Read the sentence from the story.

The villagers were soon bone tired!

What does the idiom *bone tired* mean? If you're not sure, think about how the villagers felt after rushing to put out so many fires.

7. Read the last sentences of the story.

Papa agreed. "You and Grandmother Maris both know your way around stitches!"

Think about how Papa helps Maris and what you know about parents and children. Then make an inference: How does Papa feel about his daughter as the story ends?

8. State the story's theme, or big idea about life, in your own words.

Read the adaptation of "Rip Van Winkle" by Washington Irving. Then circle the correct answer.

The Legend of Rip Van Winkle

Rip Van Winkle lived in a small village nestled in the Catskill Mountains. Rip was a kind, happy man. Everyone loved him—especially children. But his wife did not! Rip loved to fish and hunt. He liked to sit in the inn. He loved to chat with friends. But he did not like to work on his farm. His fields were full of weeds. His fences needed repair. His children ran ragged and wild. When his wife nagged him to fix things, Rip shrugged and left the house.

One day, Rip's wife was harping on all his undone work. So Rip hiked up the mountain to get away from his wife's scolding. High up, he stopped to admire the view. Suddenly, he heard someone call his name. He saw a short little man—a stranger wearing strange clothes. The stranger wore several pairs of short pants, one over the other. He seemed harmless. Rip followed the stranger to a valley. There, more people dressed like the little man were playing a game like bowling. They looked very serious as they played. The sound of the balls hitting the pins was like thunder!

The people offered Rip a drink. They all drank, in silence, a strange-tasting brew. Soon Rip felt cloudy. He fell into a deep sleep. When Rip awoke, he found himself back on the mountainside, where he had first seen the little man. Rip felt hungry. He wanted to go home to eat. But he knew his wife would be angry. He must have slept all day and all night!

Rip felt stiff as he got up and walked down to the village. When he arrived, he saw people and rows of houses that he had never seen before. And his own house looked old and empty! The roof had fallen in. His wife was nowhere to be seen. He called her name. He called his children's names. But he heard only silence.

Rip looked for familiar faces in the village. Strangers crowded around him. He asked about old friends. He learned that many of them had died. In desperation, he asked, "Does nobody here know Rip Van Winkle?"

A young woman with a child in her arms stepped up. "My father's name was Rip Van Winkle," she said. "He went up on the mountain twenty years ago. No one has heard of him since then." She also said that her mother had passed away not too long ago.

Rip cried out, "I am your father!" He told everyone about the strange little people and the bowling game. He said that he had slept not one night but twenty years. No one believed him. They thought he had lost his mind.

Rip lived out his years with his daughter and her family. He was happy because he never had to do any work. He became a village favorite again. He told his story to all who would listen. Some never believed him. Others swore it was all true, especially when they heard thunder echoing in the mountains!

1. Where and when is the story set?

   A. in a pretend city long ago

   B. in a small village in the past

   C. in a mountain town in the present

   D. in an imagined place in the future

2. Who is the main character in the story?

   A. Rip

   B. Rip's wife

   C. the oddly-dressed man

   D. Rip's grown-up daughter

3. What character trait drives most of Rip's decisions?

   A. his laziness

   B. his kind nature

   C. his love of his family

   D. his sense of adventure

4. Which event marks the plot's climax?

   A. Rip goes to the mountains to avoid work.

   B. Rip drinks something that tastes odd and falls asleep.

   C. Rip learns that he has slept for twenty years.

   D. Rip enjoys telling his story to anyone who will listen.

5. Reread the last paragraph of the story. The signal word *because* links the effect, "He was happy," to its cause. What is the cause?

   A. He lived with his daughter and her family.

   B. He never had to do any work.

   C. He became a village favorite again.

   D. He told his story to all who would listen.

6. When Rip does not return to the village for many years, what do the villagers probably conclude about him?

   A. His wife killed him.

   B. He has left the village or died.

   C. He is camping on the mountain.

   D. He is bowling with the strange little people.

7. Based on Rip's response to his wife before he went up the mountain, what can readers infer about his response to the news that his wife has died?

   A. He feels angry.

   B. He feels very sad.

   C. He feels confused.

   D. He feels a bit relieved.

8. Which theme does the story suggest?

   A. Rip should keep his silly tales to himself.

   B. People should be careful about trusting strangers.

   C. Hard-working people are successful while lazy people miss out.

   D. Adult children must care for their elderly parents.

9. When Rip tells his story, the villagers "thought he had lost his mind." Which phrase best explains what the idiom *lost his mind* means?

   A. was crazy or foolish

   B. would not tell the truth

   C. became angry or distressed

   D. could not remember anything

10. Read the sentence from the story.

When his wife nagged him to fix things, Rip shrugged and left the house.

   When people <u>nag</u> someone to do something, how do they speak?

   A. angrily

   B. firmly

   C. kindly

   D. proudly

NOTES

NOTES

NOTES

NOTES

NOTES